W0234819

Geotechnical Engineering: Soil Mechanics

Geotechnical Engineering: Soil Mechanics

Contributors

Santosh Kumar Sarkar, Paulo J.C. Favas et al.

AURIS
Reference

www.aurisreference.com

Geotechnical Engineering: Soil Mechanics

Contributors: Santosh Kumar Sarkar, Paulo J.C. Favas et al.

Published by Auris Reference Limited
www.aurisreference.com

United Kingdom

Copyright 2016
Printed in 2017 for Sale in the Indian Subcontinent

The information in this book has been obtained from highly regarded resources. The copyrights for individual articles remain with the authors, as indicated. All chapters are distributed under the terms of the Creative Commons Attribution License, which permit unrestricted use, distribution, and reproduction in any medium, provided the original author and source are credited.

Notice

Contributors, whose names have been given on the book cover, are not associated with the Publisher. The editors and the Publisher have attempted to trace the copyright holders of all material reproduced in this publication and apologise to copyright holders if permission has not been obtained. If any copyright holder has not been acknowledged, please write to us so we may rectify.

Reasonable efforts have been made to publish reliable data. The views articulated in the chapters are those of the individual contributors, and not necessarily those of the editors or the Publisher. Editors and/or the Publisher are not responsible for the accuracy of the information in the published chapters or consequences from their use. The Publisher accepts no responsibility for any damage or grievance to individual(s) or property arising out of the use of any material(s), instruction(s), methods or thoughts in the book.

Geotechnical Engineering: Soil Mechanics

ISBN: 978-1-78154-913-1

British Library Cataloguing in Publication Data
A CIP record for this book is available from the British Library

Printed in the United Kingdom

Exclusively distributed by CBS Publishers & Distributors Pvt. Ltd.

Sales & Distribution Rights only for India, Pakistan, Bangladesh, Sri Lanka, Nepal and Bhutan.This book is not to be sold outside these territories.

Contents

vi

List of Abbreviations

ASF	Alaska Satellite Facility
ANN	Artificial Neural Networks
ABNT	Brazilian Association of Technical Standards
CBR	Californian bearing ratio
CEC	cation exchange capacity
CRM	certified reference materials
C.I	confidence intervals
DC	Dynamic compaction
EDTA	Ethylenediaminetetraacetic acid
EK	Electrokinetic
EIP	Engineering index properties
EPA	Environmental Protection Agency
FEM	Finite Element Method
FRA	Forest Reclamation Approach
GP	Genetic programming
GPS	global positioning system
HCA	Hierarchical cluster analysis
HRE	Hugli (Ganges) River Estuary
INL	intermediate nodular layer
MDD	maximum dry density
MICP	Microbial induced carbonate precipitation
MDL	minimum detection limits
MCL	mottled clayey layer
NDG	Nuclear Density Gage
OMC	optimum moisture content
OM	organic matter
POPs	persistent organic pollutants
PC	principal component
PCA	Principal component analysis
PEs	Processing elements
QC	quality control
RAC	risk assessment code
SEP	sequential extraction procedure
SM	silty sands
SD	standard deviations
SAR	Synthetic Aperture Radar
TL	Thin Lamina
T.I	tolerance intervals

List of Contributors

Santosh Kumar Sarkar
Department of Marine Science, University of Calcutta, Calcutta, West Bengal, India

Paulo J.C. Favas
Department of Geology, School of Life Sciences and the Environment, University of Trás-os-Montes e Alto Douro, Vila Real, Portugal
IMAR-CMA Marine and Environmental Research Centre, Faculty of Sciences and Technology, University of Coimbra, Coimbra, Portugal

Dibyendu Rakshit
Department of Marine Science, University of Calcutta, Calcutta, West Bengal, India

K.K. Satpathy
Indira Gandhi Centre for Atomic Research, Environment and Safety Division, Kalpakkam, Tamil Nadu, India

Wisley Moreira Farias
Universidade de Brasília, Brasília-DF, Brazil

Geraldo Resende Boaventura
Universidade de Brasília, Brasília-DF, Brazil

Éder de Souza Martins
Embrapa/ Cerrados, Planaltina –DF, Brazil

Fabrício Bueno da Fonseca Cardoso
ANA-Agência Nacional de Águas –Brasília-DF, Brazil

José Camapum de Carvalho
Universidade de Brasília, Brasília-DF, Brazil

Edi Mendes Guimarães
Universidade de Brasília, Brasília-DF, Brazil

A.K.M. Azad Hossain
Department of Geology and Geological Engineering, The University of Mississippi, 120A Carrier Hall, University, MS 38677, USA

Greg Easson
Mississippi Mineral Resources Institute, The University of Mississippi, University, MS 38677, USA

Dale W. Griffin
Coastal and Marine Science Center, U.S. Geological Survey, 600 4th Street South, St. Petersburg, FL 33701, USA

Erin E. Silvestri
National Homeland Security Research Center, U.S. Environmental Protection Agency, 26 W. Martin Luther King Drive, MS NG16, Cincinnati, OH 45268, USA

Charlena Y. Bowling
National Homeland Security Research Center, U.S. Environmental Protection Agency, 26 W. Martin Luther King Drive, MS NG16, Cincinnati, OH 45268, USA

Timothy Boe
National Homeland Security Research Center, Oak Ridge Institute for Science and Education, with the U.S. Environmental Protection Agency, 109 T.W. Alexander Drive, Research Triangle Park, NC 27709, USA

David B. Smith
Denver Federal Center, U.S. Geological Survey, Box 25046, MS 973, Denver, CO 80225, USA

Tonya L. Nichols
National Homeland Security Research Center, Threat and Consequence Assessment Division, U.S. Environmental Protection Agency, Ronald Reagan Building, MC 8801RR, 1200 Pennsylvania Avenue, NW, Washington, DC 20460, USA

Mohamed A. Shahin
Department of Civil Engineering, Curtin University of Technology, Perth, WA 6845, Australia

Mark B. Jaksa
School of Civil, Environmental and Mining Engineering, University of Adelaide, Adelaide, SA 5005, Australia

Holger R. Maier
School of Civil, Environmental and Mining Engineering, University of Adelaide, Adelaide, SA 5005, Australia

Isaac A. Jeldes
Department of Civil and Environmental Engineering, University of Tennessee, Knoxville, TN 37996, USA

Eric C. Drumm
Department of Biosystems Engineering and Soil Science, University of Tennessee, Institute of Agriculture, Knoxville, TN 37996, USA

John S. Schwartz
Department of Civil and Environmental Engineering, University of Tennessee, Knoxville, TN 37996, USA

Luı́s Pinheiro Branco
L. Pinheiro Branco (&) AdFGeo - Consultores de Geotecnia, Rua Ferna~o Lopes, 157 - 48 Esq, Porto 4150-318, Portugal

Antonio Topa Gomes
A. Topa Gomes A. Silva Cardoso Department of Civil Engineering, Faculty of Engineering, University of Porto, Rua Dr. Roberto Frias, Porto 4200-465, Portugal

Anto´nio Silva Cardoso
A. Topa Gomes A. Silva Cardoso Department of Civil Engineering, Faculty of Engineering, University of Porto, Rua Dr. Roberto Frias, Porto 4200-465, Portugal

Carla Santos Pereira
C. Santos Pereira Portucalense University Infante D. Henrique, Rua Dr. Anto´nio Bernardino de Almeida, 541, Porto 4200-072, Portugal

Hamed A. Keykha
Huat Department of Civil Engineering, Faculty of Engineering, Universiti Putra Malaysia (UPM), 43400 Serdang, Selangor, Malaysia

Bujang B. K. Huat
Huat Department of Civil Engineering, Faculty of Engineering, Universiti Putra Malaysia (UPM), 43400 Serdang, Selangor, Malaysia

Afshin Asadi
A. Asadi (&) Faculty of Engineering, Housing Research Center, Universiti Putra Malaysia (UPM), 43400 Serdang, Selangor, Malaysia

Mohammed Y. Fattah
Building and Construction Engineering Department, University of Technology, Baghdad, Iraq

Hawraa H. M. al-Musawi
Civil Engineering Department, College of Engineering, University of Kufa, Kufa, Iraq

Firas A. Salman
Department of Civil Engineering, Faculty of Engineering, University of Malaya, 50603 Kuala Lumpur, Malaysia

Brendan c. O'kelly
Department of Civil, Structural and Environmental Engineering, University of Dublin, Trinity College, Dublin, 2, Ireland

Brice T Kamtchueng
Department of Earth, Life and Environmental Sciences, Graduate School of Science and Engineering for Education, University of Toyama, Gofuku 3190, Toyama 930-8555, Japan
Department of Earth and Life Sciences, Faculty of Sciences, University of Yaounde I, Yaounde, Cameroon

Vincent L Onana
Department of Earth and Life Sciences, Faculty of Sciences, University of Yaounde I, Yaounde, Cameroon

Wilson Y Fantong
Institute of Mining and Geological Researches (IRGM), Yaounde, Cameroun

Akira Ueda
epartment of Earth, Life and Environmental Sciences, Graduate School of Science and Engineering for Education, University of Toyama, Gofuku 3190, Toyama 930-8555, Japan

Roger FD Ntouala
Department of Earth and Life Sciences, Faculty of Sciences, University of Yaounde I, Yaounde, Cameroon

Michel HD Wongolo
Department of Earth and Life Sciences, Faculty of Sciences, University of Yaounde I, PO Box 812, Yaounde, Cameroon

Ghislain B Ndongo
Department of Earth and Life Sciences, Faculty of Sciences, University of Yaounde I, PO Box 812, Yaounde, Cameroon

Arnaud Ngo'o Ze
Department of Earth and Life Sciences, Faculty of Sciences, University of Yaounde I, PO Box 812, Yaounde, Cameroon

Véronique KB Kamgang
Higher Teacher's Training College (ENS), University of Yaounde I, PO Box 4110, Yaounde, Cameroun

Joseph M Ondoa
Department of Earth and Life Sciences, Faculty of Sciences, University of Yaounde I, PO Box 812, Yaounde, Cameroon

Preface

Geotechnical engineering is the branch of civil engineering concerned with the engineering behavior of earth materials. Geotechnical engineering is important in civil engineering, but also has applications in military, mining, petroleum and other engineering disciplines that are concerned with construction occurring on the surface or within the ground. Geotechnical engineering uses principles of soil mechanics and rock mechanics to investigate subsurface conditions and materials; determine the relevant physical/mechanical and chemical properties of these materials; evaluate stability of natural slopes and man-made soil deposits; assess risks posed by site conditions; design earthworks and structure foundations; and monitor site conditions, earthwork and foundation construction. Geotechnical Engineering Soil Mechanics focuses on soil engineering, updated to include the latest soil testing methodologies and technologies. First chapter aims to summarize the potentials of sequential extraction technique adopting different analytical protocols for gaining information on the mobility and dynamics of operationally determined chemical forms of heavy metals in soils and sediments. The aim of second chapter is to evaluate the behavior of a tropical soil, and their performance as liner against the flow of hydrocarbons from gasoline, by interpreting transportation according to physical and chemical parameters, as well as micromorphological aspects. Third chapter investigates the potential of multiple linear regressions and Artificial Neural Networks (ANN) based models to improve soil moisture estimation in south-eastern New Mexico using high resolution Radarsat 1 SAR imagery. Chapter four discusses anthrax and the geochemistry of soils in the contiguous United States. Fifth chapter presents a brief overview of ANN applications in geotechnical engineering, briefly provides an overview of the operation of ANN modeling, investigates the current research directions of ANNs in geotechnical engineering, and discusses some ANN modeling issues that need further attention in the future, including model robustness; transparency and knowledge extraction; extrapolation; uncertainty. The objectives of sixth chapter are to characterize the geotechnical properties of low compacted spoils on steep slopes constructed according to the FRA, and investigate the likely failure mechanisms associated to steep slopes reclaimed using the FRA. Natural variability of shear strength in a granite residual soil from Porto has been described in seventh chapter and eighth chapter presents on the electrokinetic stabilization of soft soil using carbonate-producing bacteria. The purpose of ninth chapter is to investigate the effect of dynamic compaction process on the behavior of gypseous soils. Compression and consolidation anisotropy of some soft soils are described in tenth chapter. Geotechnical, chemical and mineralogical evaluation of lateritic soils in humid tropical area are presented in last chapter.

Chapter 1

GEOCHEMICAL SPECIATION AND RISK ASSESSMENT OF HEAVY METALS IN SOILS AND SEDIMENTS

Santosh Kumar Sarkar[1], Paulo J.C. Favas[2, 3], Dibyendu Rakshit[1] and K.K. Satpathy[4]

[1]Department of Marine Science, University of Calcutta, Calcutta, West Bengal, India

[2]Department of Geology, School of Life Sciences and the Environment, University of Trás-os-Montes e Alto Douro, Vila Real, Portugal

[3]IMAR-CMA Marine and Environmental Research Centre, Faculty of Sciences and Technology, University of Coimbra, Coimbra, Portugal

[4]Indira Gandhi Centre for Atomic Research, Environment and Safety Division, Kalpakkam, Tamil Nadu, India

INTRODUCTION

Heavy metal pollution is a serious and widely environmental problem due to the persistent and non-biodegradable properties of these contaminants. Sediments serve as the ultimate sink of heavy metals in the marine environment and they play an important role in the transport and storage of potentially hazardous metals. They are introduced into the aquatic system as a result of weathering of soil and rocks, from volcanic eruptions and from a variety of human activities involving mining, dredging, processing and use of metals and/or substances containing metal contaminants. Heavy metals entering natural water become part of the water-sediment system and their distribution processes are controlled by a dynamic set of physicochemical interactions and equilibria. The properties of metals in soils and sediments depend on the physiochemical form in which they occur [1]. Heavy metals are distributed throughout soil and sediment components and associated with them in various ways, including adsorption, ion exchange, precipitation and complexation and so on [2]. Changes in environmental conditions, such as temperature, pH, redox potential and organic ligand concentrations, can cause metals to be released from solid to liquid phase and sometimes cause contamination of surrounding waters in aquatic systems [3]. They are not permanently fixed by soil or sediment. Therefore, it cannot provide sufficient information about mobility, bioavailability and toxicity of metals if their total contents are studied alone.

Natural and anthropogenic activities have the capacity to cause changes in environment conditions, such as acidification, redox potential, or organic ligand concentrations, which can remobilize contaminated soils and sediments releasing the elements from soils and sediments and pore water to the water column resulting contamination of surrounding waters. Daily tidal currents, wind energies, and storms in coastal and estuarine systems can cause periodical remobilization of surface sediments [4]. More turbulent conditions, such as seasonal flooding or storms, or bioturbation, due to feeding and movement of benthic organisms, can expose anoxic sediments to oxidant conditions. In addition, activities such as dredging result in major sediment disturbances, leading to changes in chemical properties of sediment [5].

The remediation of heavy metal pollution is often problematic due to their persistence and non-degradability in the environment. As a sink and source, soils and sediments constitute a reservoir of bioavailable heavy metals and play a significant role in the remobilization of contaminants in the aquatic systems under favorable conditions. Such potential of sediment for being a sink as well as a source of contaminant can make sediment chemistry and toxicity key components of the quality of aquatic system. Much concern has been focused on the investigation of the total element contents in soils and sediments. However, it cannot provide sufficient information about mobility, bioavailability and toxicity of elements and thus may not be able to provide information about the exact dimension of pollution. The data on total contents of metals are quite insufficient to estimate the possible risk of remobilization of total metals under changing environmental conditions and potential uptake of liberated metals by biota and thus the determination of different fractions assume great importance. This has been described as "speciation" [6]. Since each form have different bioavailability and toxicity, the environmentalists are rightly concerned about the exact forms of metal present in the aquatic environment.

The concept of speciation dates back to 1954 when Goldberg introduced the concept of speciation to improve the understanding of the biogeochemical cycling of trace elements in seawater. Kinetic and thermodynamic information together with the analytical data made it possible to differentiate between oxidized versus reduced, complexed or chelated versus free metal ions in solution and dissolved between particulate species. Florence [7] has defined the term speciation analysis as the determination of the individual physicochemical forms of the element, which together make up its total concentration in a sample. According to Lung [8], speciation analysis involves the use of analytical methods that can provide information about the physicochemical forms of the elements. Schroeder [9] distinguishes physical speciation, which involves differentiation of the physical size or the physical properties of the metal, and chemical speciation, which entails differentiation among the various chemical forms. The main objective of measuring metal species relates to their relative toxicities to aquatic biota. The second and long term aim of speciation studies is to advance an understanding of metal interactions between water and bed sediments in an aquatic ecosystem. In the last decade researchers have followed different sequential extraction techniques for the fractionation of metals in sediments of different river systems. Rauret et al. [10] studied the speciation of copper and lead in the sediments of River Tenes (Spain) while Pardo et al. [11] studies the speciation of zinc, cadmium, lead, copper, nickel and cobalt in the sediments of Pisuerga River, Spain, in order to establish the extent to which these are polluted and their capacity to remobilization. Jardo and Nickless [12] investigated the chemical association of zinc, cadmium, lead and copper in soils and sediments of England and Wales. In most samples, these four metals were associated with all the chemical fractions. Tessier et al. [13] studied speciation of cadmium, cobalt, copper, nickel, lead, zinc, iron and manganese in water and sediments of St. Fransois River, Quebec, Canada. Elsokkary and Muller [14] studied speciation of chromium, nickel, lead and cadmium in the sediments of Nile River, Egypt, reporting that a high proportion of chromium, nickel and lead is bound to organic material and sulphides, while cadmium is bound to carbonate fraction. Ure [15] and Rauret [16] have reviewed the chemical extraction procedures used for heavy metal determinations in contaminated soils and sediments. Owing to the need for validation of extraction schemes, the EC Measurement and Testing Programme (formerly BCR) has organized a project for improving the quality of determinations of extractable heavy metals, where development and validation of extraction procedures has been discussed [17, 18].

The present article aims to summarize the potentials of sequential extraction technique adopting different analytical protocols for gaining information on the mobility and dynamics of operationally determined chemical forms of heavy metals in soils and sediments. The BCR (Community Bureau of Reference, now superseded by the Standards, Measurement and Testing Programme of the European Community) procedure has been illustrated considering the case study of Ganges (Hugli) River Estuary and adjacent Indian Sundarban mangrove wetland (a UNESCO World Heritage Site), northeastern part of the Bay of Bengal. In addition, the authors also evaluate the modified BCR sequential extraction technique as devised by various scientists, the risk assessment code (RAC) as well as assessment of toxicity comparing with sediment quality guidelines. The RAC classification is based on the strength of the bond between the metals and the different geochemical fractions in sediments or soils and the ability of metals to be released and enter into the food chain.

SEQUENTIAL EXTRACTION: MERITS AND DEMERITS

The sequential extraction provides more or less detailed information concerning the origin, mode of occurrence, biological and physicochemical availabilities, mobilization and transport of heavy metals. The procedure stimulates the mobilization and retention of these species in the natural environment using changes in environmental condition such as pH, redox potential and degradation of organic matter [16]. A series of reagents is applied to the sample, increasing the strength of the extraction at each step, in order to dissolve the trace metal present in different sediment phases. The extractants are inert electrolytes, weak acids, reducing agent, oxidizing agents and strong mineral acids [19].

The 3-stage sequential extraction procedure proposed by the European Community Bureau of Reference (BCR) was developed in an attempt to standardize the various schemes described in the literature [2, 20,21], since the use of different procedures, varying in the number of steps, types of reagents and extraction condition. Hindered comparison of results obtained in the many studies of heavy metals chemical fractionation in environmental samples [22].

The BCR methods has been widely adapted by various authors, and applied to a range of type of solid sample including fresh water sediment [23-25], salt water sediment [26-28], sewage sludge and particulate matter [29-31]. This scheme enables us to associate the meals with one of the following four geochemical phases:

I.   Acid-soluble phase: This phase is made up of exchangeable metals and others bound to carbonates that are able to pass easily into the water column, for example, when the pH drops. It is the fraction with the most labile bond to the soil/sediment and, therefore, the most dangerous for the environment.

II.   Reducible phase: This phase consists of metals bound to iron and manganese oxides that can be released if the sediment changes from the oxic to the anoxic state, which could be caused, for example, by the activity of microorganisms present in the soils/sediments.

III.   Oxidisable phase: This shows the amount of metal bound to organic matter and sulphides, which can be released under oxidizing conditions. Such conditions can occur, for example, if the sediment is resuspended (by dredging, currents, flooding, tides, etc.) and the sediment participles come into contact with oxygen-rich water.

IV.   Residual phase : Lithogenous and inert (Non-bioavailable).

The heavy metals in the soils and sediments are bound to different fractions with different strengths, the value can, therefore, give a clear indication of soil and sediment reactivity, which in turn assess the risk connected with the presence of heavy metals in a terrestrial or aquatic environment. The rationale of the sequential extraction procedure is that each successive reagent dissolves a different component, which can content heavy metals within their crystalline structures. Under natural conditions, metals in minerals are unlikely to experience significant release over the time frames of interest [32, 33].

ANALYTICAL PROTOCOLS FOR SEQUENTIAL EXTRACTION

In recent years a great number of papers have been published on various analytical techniques proposed for the fractionation analysis of trace elements in various environmental samples (soils, sediments, etc.). An approach that has been found to be preferable is the fractionation of heavy metal into operationally defined forms under the sequential action of different extractants [2]. Selective extractants, used in sequential extraction procedures, are aimed at the simulation of natural conditions whereby metals associated with certain soil (sediment) components can be released. For example, changes in the ionic composition affecting adsorption–desorption reactions or a decrease in pH may lead to the release of metals, retained on a matrix by weak electrostatic interactions or co-precipitated with carbonates ("exchangeable" and "acid soluble" forms). Decreasing the redox potential can result in dissolution of oxides, unstable under reducing conditions, and liberation of scavenged metals ("reducible" forms). Changes in oxidizing conditions may cause the degradation of organic matter and release of complexed metals ("oxidizable" forms). Finally, the destruction of primary and secondary mineral lattice releases heavy metal retained within the crystal structure, e.g., due to isomorphous substitution ("residual" forms) [2]. The nominal "forms" determined by operational fractionation can help to estimate the amounts of total metals in different reservoirs which could be mobilized under changes in the chemical properties of the soil [34]. Since the 1970s a considerable number of extraction procedures have been proposed for determining the forms of heavy metal [2, 35-39]. Most of these procedures are based on the scheme of Tessier et al. [2]. Although most of the extracting reagents were originally used in the chemical analysis of soils, the procedures proposed have been tested on a wide variety of contaminated environmental samples—sediments, road dust, sewage sludge, etc.

Sequential extraction can be useful to have an operational classification of metals in different geochemical fractions [2] which is the most reliable criteria to quantify the potential effect of soil/sediment contamination by heavy metals. This can provide information about the identification of the main binding sites, the strength of element binding to the particulates and the phase associations of trace elements in soil/sediment. Following this basic scheme, some modified procedures with different sequences of reagents or operational conditions have been developed [40-43]. Considering the diversity of procedures and the lack of uniformity in different protocols, a European Community Bureau of Reference (BCR, now the European Community Standards Measurement and Testing Program) method was proposed [6] and was applied by a large group of researchers [31, 44-47]. In this study, we followed the sequential extraction procedure proposed by the European Union's Standards, Measurements and Testing program [3].

MODIFIED BCR SEQUENTIAL EXTRACTION PROCESS

As discussed above it is evident that sequential extraction provides valuable information regarding identification of main binding site, the strength of the element binding to the particulates and the phase associations of heavy metals in sediments. However, various complicated sequential extraction procedures were experimented to provide more detailed information regarding different metal phase associations [2, 48, 49]. A wide range of techniques is available whereby various extraction reagents and experimental conditions are used. These techniques involve a 5-step [2], 4-step (BCR, Bureau Commune de Reference of the European Commission), 6-step [50] and 7-step [51, 52] extraction, and are thus becoming popular methods to be used for sequential extraction [53, 54]. Following this basic scheme, some modified procedures with different sequences of reagents or operational conditions have been developed [40-43].

Several sophisticated instruments have been used for the determination of heavy metals contents in marine environments. These include; flame AAS [55, 56], atomic fluorescence spectrometry [57], anodic stripping voltametry [58, 59], ICP-AES [60] and ICP-MS [61, 62].

Heavy metal mobility and bioavailability depend strongly on their chemical and mineralogical forms in which they occur [63]. Several speciation studies have been conducted to determine study different forms of heavy metals rather their total metal content. These studies reveal the level of bioavailability of metals in harbour sediments and also confirm that sediments are indicators of heavy metal pollution in marine environment [64-67].

Since the early 1980s and 1990s sequential extraction methodology has been developed to determine speciation of metals in sediments [2, 68] due to the fact that the total concentration of metals often does not accurately represent their characteristics and toxicity. In order to overcome the above mentioned obstacles it is helpful to evaluate the individual fractions of the metals to fully understand their actual and potential environmental effects [2]. To date, strong acid digestion is used often for the determination of total heavy metals in the sediments. However, this method can be misleading when assessing environmental effects due to the potential for an overestimation of exposure risk. Moreover, in order to determine the mobility of heavy metals in sediments, various sequential extraction procedures have been developed [69-71].

Among a range of available techniques using various extraction reagents and experimental conditions to investigate the distribution of heavy metals in sediments and soils, the 5-step Tessier et al. [2] and the 6-step extraction method, Kersten and Fronstier [50] were mostly widely used. Following these two basic schemes, some modified procedures with different sequences of reagents or experimental conditions have been developed [40-43]. Considering the diversity of procedures and lack of uniformity in different protocols, a BCR, Bureau Commun de Recherche (now called the European Community (EC) Standards Measurement and Testing Programme) method was proposed [6]. It harmonized differential extraction schemes for sediment analysis. The method has been validated using a sediment certified reference material BCR-701 with certified and indicative extractable concentration of Cd, Cr, Cu, Ni, Pb and Zn [72]. This method was applied and accepted by a large group of specialists [31, 44, 45, 47,73, 74] despite some shortcoming in the sequential extraction steps [75, 76].

Wang et al. [77] used a modified Tessier sequential extraction method to investigate the distribution and speciation of Cd, Cu, Pb, Fe, and Mn in the shallow sediments of Jinzhou Bay, Northeast China. This site was heavily contaminated by nonferrous smelting activities. They found out that the concentrations of Cd, Cu and Pb in sediments was to be 100, 73, 13 and 7 times, respectively, higher than the National guidelines (GB 18668-2002). The sequential extraction tests revealed that 39%-61% of Cd was found in exchangeable fractions. This shows that Cd in the sediments posed a high risk to the local environment. Copper and Pb were found to be at moderate risk levels. According to the relationships between percentage of metal speciation and total metal concentration, it was concluded that the distributions of Cd, Cu and Pb in some geochemical fractions were dynamic in the process of pollutants migration and stability of metals in marine sediments from Jinzhor Bay decrease in the order Pb>Cu>Cd.

Yuan et al. [78] applied BCR-sequential extraction protocol to obtain metal distribution patterns in marine sediments from the East China Sea. The results showed that both the total contents and the most dangerous non-residual fractions of Cd and Pb were extremely high. More than 90% of the total concentration of V, Cr, Mo and Sn existed in the residual fraction while more than 60% of Fe, Co, Ni, Cu, and Zn were mainly present in the residual fraction. Manganese, Pb, and Cd were dominantly present in the non-residual fractions in the top sediments.

Jones and Turki [79] worked on distribution and speciation of heavy metals in surface sediments from the Tees estuary, North East England. Tessier et al. [2] metal speciation scheme modified by Ajay and van Loon [80] was used for the study. They observed out that the sediments were largely organic-rich clayey silts in which metal concentrations exceed background levels, and which attain peak values in the upper and middle reaches of the estuary. Chromium, Pb and Zn were associated with the reducible, residual, and oxidizable fractions. Cobalt and Ni were not highly enriched while Cu is associated with the oxidizable and residual fractions. Cadmium is associated with the exchangeable fractions.

Pempkowlak et al. [81] investigated the speciation of heavy metals in sediments and their bioaccumulation by mussels. They used a 4-step sequential extraction procedure adapted from Forstner and Watmann [82]. Their investigation which was characterized by varying metal bioavailability was aimed at revealing differences in the accumulation pattern of heavy metals in mussel inhabiting that inhabit in sediments. The bioavailabilities of metals were measured using the contents of metals adsorbed to sediments and associated with Fe and Mn hydroxides. The biovailable fraction of heavy metals contents in sediments collected from Spitsbergen represented a small proportion (0.37% adsorbed metals and 0.11%, are associated with metals hydroxides). It was also revealed that the percentages of metals adsorbed and bound to hydroxides of the sediments ranged from 1 to 46% and 1 to 13%, respectively.

Wepener and Vermeulen [66] investigated on the concentration and bioavailability of selected metals in sediments of Richards Bay harbor, South Africa. Sequential extraction of sediments was carried out according to Tessier et al. [2] method. The following metals were investigated: Al, Cr, Fe, Mn, and Zn. Their studies revealed that metals concentrations in sediments samples varied only slightly between seasons, but showed significant spatial variation, which was significantly correlated to sediment particle size composition.

Highest metal concentration was recorded in sites with substrates dominated by fine mud. Manganese and Zn had more than 50% of this concentration in reducible fraction while more than 70% of the Cr was associated with the inert fractions and the concentration recorded at some sites were still above action levels when considering only the bioavailable fractions. They also concluded that the concentration of Zn recorded was not elevated their results were compared with the historic data.

Coung and Obbard [54] used a modified 3-step sequential extraction procedure to investigate metal speciation in coastal marine sediments from Singapore as described by the European Community Bureau of Reference (ECBR). Highest percentages of Cr, Ni, and Pb were found in residual fractions in both Kranji (78.9%, 54.7% and 55.9% respectively) and Pulang Tokong (82.8%, 77.3% and 62.2% respectively). This means that these metals were strongly bound to sediments. In sediments from Kranji, the mobility order of heavy metals studied were Cd>Ni>Zn>Cu>Pb>Cr while sediments from Pulan Tekong showed the same order for Cd, Ni, Pb and Cr, but had a reverse order for Cu and Zn (Cu>Zn). The sum of the 4-steps (acid soluble + reducible + oxidizable + residual) was in good agreement with the total metal content, which confirmed the accuracy of the microwave extraction procedure in conjunction with the GFASS analytical method.

Fedotov et al. [83] applied a modified technique for accelerated fractionation of heavy metals in contaminated soils and sediments using rotating coiled columns. Rotating coiled columns (RCC) is valuable for the continuous-flow sequential extraction and can be successfully applied to the dynamic leaching of heavy metals from soil and sediments. This is a fluoroplastic or steel coil wound around a rigid cylindrical drum, which revolves about its axis and, at the same time, revolves around the central axis of the device called planet centrifuge. The stationary (liquid, solid, or heterogeneous) phase is retained in the column because of the centrifugal force field, and the mobile liquid phase is continuously pumped through the column. A solid sample was retrieved in the rotating column as the stationary phase under the action of centrifugal forces while different elements (aqueous solution of complexing reagents, mineral salts and acids) were continuously pumped through. This procedure developed is time saving and requires only 4-5 hr instead of the several days needed for individual sequential extraction. Losses of solid sample are minimal. Further studies are needed to better estimate the reproducibility of the technique.

Nemati et al. [84] used a modified BCR sequential extraction procedure (SEP) in combination with ICP-MS to obtain the metal distribution patterns in different depths of sediments from Sungai Buloh, Selangor, Malaysia. The results showed that heavy metal contaminations at Sungai Buloh River sediments were more severe than at other sampling sites, especially for Zn, Cu, Ni and Pb. Nevertheless, the element concentrations from top to bottom layers decreased predominantly.

Mossop et al. [85] compared of original and modified BCR sequential extraction procedures for the fractionation of Cu, Fe, Pb, Mn and Zn in soils and sediments. The procedures were applied to five soil and sediment substrates: a sewage sludge-amended soil, two different industrially contaminated soils, river sediment and intertidal sediment. Extractable Fe and Mn concentrations were measured to assess the effects of the procedural modifications on dissolution of the reducible matrix components. Statistical analyses (two-tailed t-tests at 95% confidence interval) indicated that recovery of Fe in step 2 was not markedly enhanced when the intermediate protocol was used. However, significantly greater amounts were isolated with the revised BCR scheme than with the original procedure. Copper behaved similarly to Fe. Lead recoveries were increased by use of both modified protocols, with the greatest effect occurring for the revised BCR extraction. In contrast, Mn and Zn extraction did not vary markedly between procedures. The work indicates that the revised BCR sequential extraction proves better attack on the Fe-based components of the reducible matrix for a wide range of soils and sediments.

SEQUENTIAL EXTRACTION OF METALS IN SEDIMENTS OF THE HUGLI RIVER ESTUARY AND INDIAN SUNDAR-BAN WETLAND: A CASE STUDY

Materials and Methods

Sample Collection and Sediment Quality Analysis

The delta region formed by Hugli (Ganges) River Estuary (HRE) and is famous for its luxuriant mangrove vegetation, known as Sundarban wetland, acclaimed as UNESCO World Heritage Site for its capacity of sustaining an excellent biodiversity. The wetland is characterized by a complex network of tidal creeks, which surrounds hundreds of tidal islands exposed to different elevations at high and low semi-diurnal tides. This is one of the most sensitive and vulnerable ecosystems in the world and suffers from environmental degradation due to rapid human settlement, tourism and port activities, operation of mechanized boats, deforestation, and increasing agricultural and aquaculture practices. The ongoing degradation is also related to huge siltation, flooding, storm runoff, atmospheric deposition, and other stresses resulting in changes in water quality, depletion of fishery resources, choking of river mouth and inlets, and overall loss of biodiversity. Moreover, the rapid economic development in this deltaic region has caused highly dense areas of human activity and led to serious contamination including heavy metals and persistent organic pollutants (POPs).

Nine sampling sites, namely Barrackpur (S_1), Dakshineswar (S_2), Babughat (S_3), Budge budge (S_4), Ulubaria (S_5), Diamond Harbor (S_6), Frezergunge (S_7), Gangasagar (S_8), and Haribhanga (S_9) were selected considering the existence of typical sediment dispersal patterns along the drainage network systems (as shown in Figure 1) and their position was fixed by a global positioning system (GPS). The stations are representative of the variable environmental and energy regimes that cover a wide range of substrate behavior, wave–tide climate, and intensity of bioturbation (animal–sediment interaction), geomorphological–hydrodynamic regimes and distances from the sea (Bay of Bengal). The sites are exposed to a variable level of heavy metal contamination mainly from anthropogenic sources as mentioned earlier. Six sampling sites (S_1 to S_6) have been chosen along the lower stretch of Hugli River Estuary, while residual three sites (S_7 to S_9) were taken into account in the coastal regions of Sundarban wetland. All sampling sites together with the main stresses to which they are subjected are presented inTable 1.

During winter months (January–March 2009) surface sediment samples weighting 10 g were randomly collected in triplicate from the top 3–5 cm of the surface at each sampling site during low tide using a grab sampler, pooled and thoroughly mixed. Immediately after collection, the samples were placed in sterilized plastic bags in the ice box and transported to the laboratory. Samples were oven dried at 50°C, most gently disaggregated, transferred into precleaned inert polypropylene bags and stored in deep freeze prior to analyses. Each sample was divided into two aliquots: one unsieved (for the determination of sediment quality parameters) and the other sieved through 63 μm metallic sieves (for elemental analyses). Organic carbon content was determined following a rapid titration method [86] and pH with the help of a deluxe pH meter (model no. 101E) using combination glass electrode manufactured by M.S. Electronics Pvt. Ltd. (India). Mechanical analyses of sediment were done by sieving in a Ro-Tap Shaker manufactured by W.S. Tyler Company, Cleveland, Ohio.

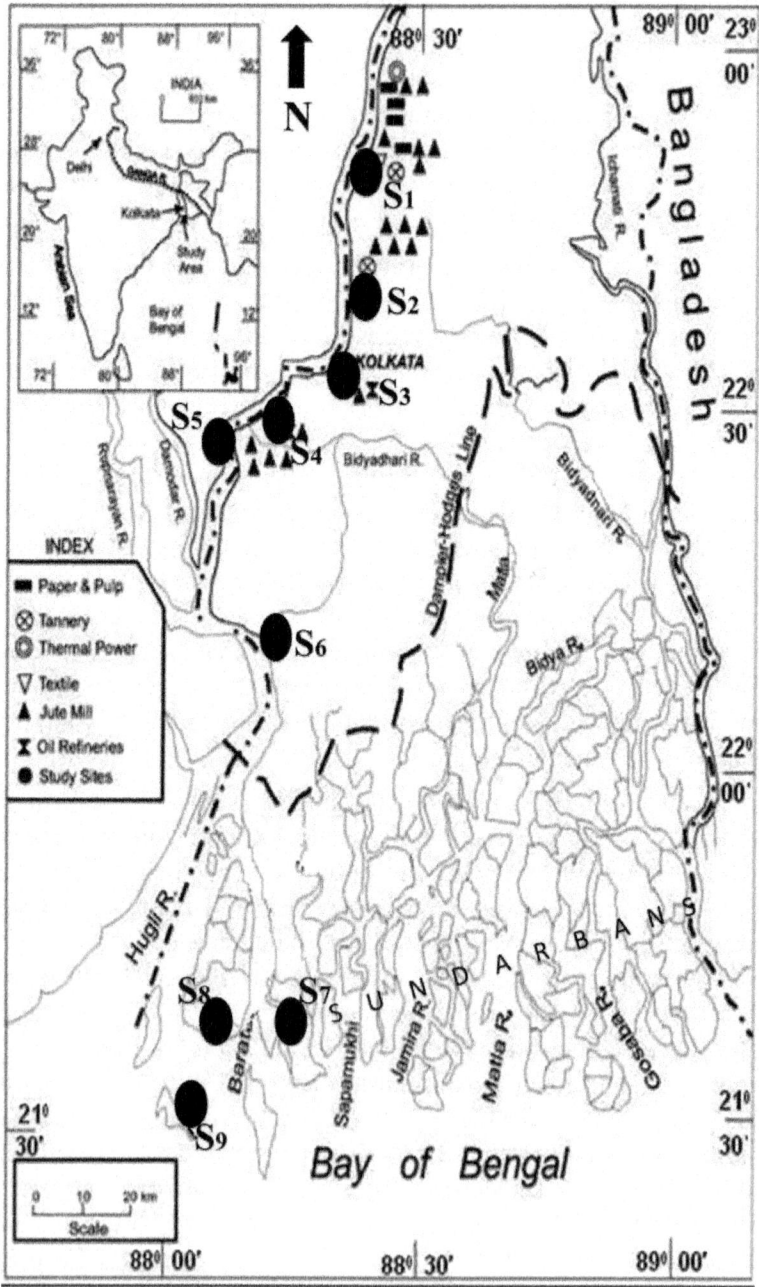

Figure 1: Map showing the location of the monitoring sites (S_1 to S_9) covering Hugli River Estuary and Sundarban mangrove wetland along with the location of the major industries.

Table 1: Details of the nine sampling sites and the main stresses to which they are subjected

Station number	Site	Main stresses
$s1$	Barrackpur	Industrial effluents, domestic sewage disposal, boating, bathing.
$s2$	Dakshineswar	Industrial and domestic effluents, boating, bathing, idol immersion site.
$s3$	Babughat	Power plant discharges, domestic sewage, boating, idol immersion site.
$s4$	Budge budge	Domestic and industrial effluents, bathing, boating.
$s5$	Ulubaria	Domestic and industrial effluents
$s6$	Diamond Harbour	Boating, recreational activities, bathing, fishing, jetties for fishing trawlers
$s7$	Frezergunge	Tourist activities, ferry services, fishing
$s8$	Gangasagar	Boating, tourist activities, dredging, fishing, agricultural, domestic and aquaculture practices
$s9$	Haribhanga	Boating, fishing and ferrying

Analytical Procedure

To determinate the total element concentration, sediment samples were digested in polytetrafluoroethylene vessels with aqua regia (HCl/HNO_3, 3:1) and HF neutralized with H_3BO_3 in a 650 W microwave oven (CEM MDS 2000) with a program consisting of a 20-min ramp and a 30-min hold at 100% power in pressure and temperature controlled conditions (150 psi and 180°C). The digested samples were filtered, transferred to polyethylene containers and stored at +4°C until analysis. All reagents were Suprapur® grade (Merck). Reagent blank was processed with the samples and did not show any significant contamination. Accuracy of the procedure was checked using two different certified reference materials (CRM): MESS-2 and PACS-2, which are both marine sediments certified by the National Research Council of Canada for the element content. The MESS-2 recovery ranged between 91% and 116% for all the elements (Table 2). Precision, calculated as relative standard deviation (RSD%), resulted always lower than 5%.

Table 2: Results of certified reference materials MESS-2 and PACS-2 as well as the observed values All the values are expressed in μg/g of dry weight. MESS-2 and PACS-2 recovery rates are also reported

	Al	As	Cd	Co	Cr	Cu	Fe	Mn	Ni	Pb	Zn
Found MESS-2	86,613± 17,773	24.0± 2.4	0.230± 0.010	15.6± 1.5	112± 12	38.4± 6.1	47,385± 3,668	372± 42	54.1± 3.8	20.0± 1.4	170± 12
Found PACS-2	70,190± 3,784	29.2± 1.3	1.59± 0.80	12.8± 0.6	94.9± 4.4	307± 22	46,630± 1,411	465± 23	44.3± 2.5	184± 10	398± 16
Certified MESS-2	85,698± 2,600	20.7± 0.8	0.240± 0.010	13.8± 1.4	106± 8	39.3± 2.0	43,504± 2,266	365± 21	49.3± 1.8	21.9± 1.2	172± 16
Certified PACS-2	66,125± 3,184	26.2± 1.5	2.11± 0.15	11.5± 0.3	90.7± 4.6	310± 12	43,738± 585	440± 19	39.5± 2.3	183±8	364± 23
Recovery MESS-2	101%	116%	95.8%	113%	106%	97.8%	109%	102%	110%	91.2%	99.0%
Recovery PACS-2	106%	111%	75.4%	111%	105%	99.1%	107%	106%	112%	100%	109%

In this study, we followed the sequential extraction procedure proposed by the European Union's Standards, Measurements and Testing program [3]. Selective extraction is based on the procedure used by Tessier et al. [2] with improvements made according to the BCR, which examined and finally eliminated irreproducibility sources. It is made up of three steps, which dissolve the following phases, respectively: exchangeable and bound to carbonate, bound to Fe and Mn oxides and hydroxides, bound to organic matter and sulphides. Exchangeable and bound to carbonate phase (phase 1) is extracted with 0.11 M acetic acid, while the fraction bound to Fe–Mn oxides (phase 2) with 0.5 M hydroxylamine hydrochloride, adjusted to pH 2 with nitric acid (65%). The phase bound to organic and sulphides (phase 3) is extracted with 8.8 M hydrogen peroxide (stabilized at a pH included between 2 and 3), treated at 80°C in a microwave oven using a program consisting of a 30-min ramp and a 60-min hold at 50% power in pressure and temperature controlled conditions (80 psi and 85°C), and 2 M ammonium acetate adjusted to pH 2 with nitric acid (65%). Each extraction was carried out overnight (16 h) at room temperature. All the reagents employed were Tracepur® grade (Merck Eurolab, Italy). After each extraction, the samples were separated from the aqueous phase by centrifuging at 4, 000 rpm for 15 min. The sediments were washed with Milli-Q water and centrifuged again. The wash water was added to supernatants. The element content of the residual phase was obtained from the difference between the total content and the sum of phases 1, 2 and 3, according to Ianni et al. [37, 38], Ramirez et al. [39], and Mester et al. [27]. Sequential extraction reagent blanks showed no detectable contamination. Accuracy of the procedure was checked with BCR-701 (SM&T). The recovery rates for trace elements in the standard reference material ranged between 77%

and 118% (Table 3). Precision, calculated as RSD%, resulted generally lower than 5%, except As and Cr in the phase 1 (~20%).

Table 3: Results of certified reference materials BCR-701 as well as the observed values (expressed in μg/g of dry weight) together with recovery rates for each step. n.a.= not available

	Al	As	Cd	Co	Cr	Cu	Fe	Mn	Ni	Pb	Zn
Found BCR-701 step 1	198±1	2.57± 0.28	6.09± 0.09	2.06± 0.08	2.41± 0.51	47.7± 1.7	43.8± 5.8	180±1	14.5± 0.3	3.38± 0.35	185±4
Found BCR-701 step 2	3,451± 46	16.5± 0.3	3.37± 0.08	3.22± 0.03	39.2± 0.4	100±2	7,042± 106	128±3	24.5± 0.4	111±2	102±1
Found BCR-701 step 3	1,912± 74	3.09± 0.20	0.28± 0.01	1.86± 0.17	169±4	64.8± 1.5	1,147± 56	31.9± 2.6	17.4± 1.7	7.15± 0.12	58.4± 5.0
Certified BCR-701 step 1	n.a.	n.a.	7.34± 0.35	n.a.	2.26± 0.16	49.3± 1.7	n.a.	n.a.	15.4± 0.9	3.18± 0.21	205±6
Certified BCR-701 step 2	n.a.	n.a.	3.77± 0.28	n.a.	45.7±2	124±3	n.a.	n.a.	26.6± 1.3	126±3	114±5
Certified BCR-701 step 3	n.a.	n.a.	0.27± 0.06	n.a.	143±7	55.2± 4.0	n.a.	n.a.	15.3± 0.9	9.3±2.0	54.2± 2.0
Recovery step 1	n.a.	n.a.	83.0%	n.a.	107%	96.8%	n.a.	n.a.	94.4%	106%	90.3%
Recovery step 2	n.a.	n.a.	89.4%	n.a.	85.9%	80.6%	n.a.	n.a.	92.3%	87.9%	89.4%
Recovery step 3	n.a.	n.a.	104%	n.a.	118%	117%	n.a.	n.a.	114%	76.9%	108%

The elemental concentrations were determined with an inductively coupled plasma atomic emission spectrometer Vista Pro (Varian), with the internal standard method. Cadmium was determined by electrothermal atomization atomic absorption spectrometry. A Varian Spectra A300 spectrometer with Zeeman effect background correction and autosampler Varian Model 96 was used employing the standard addition method for calibration. All the metal analyses were performed at the Department of Chemistry and Industrial Chemistry of the University of Genoa (Genoa, Italy).

Statistical Analyses

Principal component analysis (PCA) was used to characterize the metal composition in sediments, and cluster analysis was used for grouping the sampling stations. Principal component analysis (PCA) is a multivariate statistical technique used for data reduction and for deciphering patterns within large sets of data. With PCA, a large data matrix is reduced to two smaller ones that consist of principal component (PC) scores and loadings. PC loadings are eigenvectors of the correlation or covariance matrix depending on which is used for the analysis. The PC scores contain information on all of the variables combined into a single number, with the loadings indicating the relative contribution of each variable to that score [87]. Hierarchical cluster analysis (HCA) characterizes similarities among samples by examining interpoint distances representing all possible sample pairs in high-dimensional space. The sample similarities are represented on two dimensional diagrams call dendrograms [88]. All statistical analyses were performed using the computer software STATISTICA (StatSoft, Inc. 2001).

Results and Discussion

Sediment Geochemistry

Table 4 shows values of pH; organic carbon (%); and percentage of sand, silt, and clay in sediments of the nine sampling sites. Organic carbon values, ranging from 0.22% (in station S_8) to 1.02% (in station S_2), are low in comparison with values found in sediments from other Indian coastal areas, such as Gulf of Mannar [89], Cochin [90], and Muthupet mangroves [91]. The low organic carbon values might be related with the poor absorbability of organics on negatively charged quartz grains, which predominate in sediments in this estuarine environment [92]. In addition, the constant flushing activity by tides along with the impact of waves can support the low percentage of organic carbon in the sediments. The sediments of the studied stations are characterized by slightly basic pH (7.50–8.36) with maximum values recorded in the stations closest to the sea (stations S_6, S_8, and S_9) and minimum in station S_7.

Table 4: Geographical position, physicochemical and textural properties of sediment samples of 9 sampling sites

Stations	Latitude and Longitude	Salinity	pH	Organic carbon (%)	Sand (%)	Silt (%)	Clay (%)
S1	22°43′ 16″ N 88°21′ 20″ E	0	7.86	0.35	4	87.1	8.9
S2	22°39′ 17″ N 88°12′ 25″ E	0	7.80	1.02	1	76.5	22.5
S3	22°33′ 53″ N 88°20′ 19″ E	0	7.90	0.52	2.24	41.97	55.79
S4	22°30′ 10″ N 88°11′ 48″ E	0–2.5	7.60	0.74	18.25	47.42	34.33

S5	22°28′ 06″ N 88°06′ 54″ E	0–1	7.90	0.91	16.7	69.6	13.7
S6	22°11′ 14″ N 88°11′ 15″ E	0–5.6	8.36	0.56	3.15	41.13	55.71
S7	21°34′ 44″ N 88°15′ 03″ E	30–34.3	7.50	0.36	98.02	0.18	0
S8	21°38′ 15″ N 88°03′ 53″ E	32–35	8.14	0.22	32.85	58.45	8.7
S9	21°34′ 20″ N 88°01′ 25″ E	35	8.10	0.46	39.3	44.25	16.45

These were different from the low pH values in most of the mangrove swamps in Hong Kong [93], where sediments were not frequently flooded by the tide and become acidic in reducible conditions. With respect to texture, the sediment samples show a variable admixture of sand, silt, and clay. Clay fractions dominate in low-energy areas of suspensional deposits. On the contrary, silt, and sand dominates where the energy level is high. Sediments from station S_7 contain higher percentage of sand (98%) compared to the others, while sediments from S_1, S_2, S_5, and S_8 contains higher percentage of silt (more than 50%) compared to the others. A variable mixture of sand, silt, and clay is present in the other stations and reflect a variable amount of erosion and deposition.

Total Element Concentrations

Total element concentrations in the investigated stations varied in a narrow range of values (Table 5)and were comparable with data obtained for other Indian coastal areas [94, 95]. Datta and Subramanian [96] found very similar trace element concentrations throughout the Bengal Basin, where anthropogenic perturbation is low and river channel may receive a several centimeter-thick sediment layer in a single event during peak flow, preventing to bear the signature of an accumulation of trace elements. The highest concentrations for As, Cu, Fe, Mn, and Ni were measured at station S_9 while for Cd and Pb at station S_3, close to Calcutta city (about 4.5 million residents, but about 14.2 million including suburbs). An anthropogenic input from vehicular traffic and in-dustrial activities may cause high Cd and Pb con-centrations measured in samples collected in the Calcutta urban area. The lowest element concentrations were found at station S_5. Very low (close to the detection limit) Cd concentration was found in the coastal stations (S_7, S_8, and S_9).

Table 5: Total element concentrations (μg/g) in sediments of 9 sampling sites (instrumental precision, calculated as RSD%, resulted lower than 5% for each element in all samples)

Stations	Al	As	Cd	Co	Cr	Cu	Fe	Mn	Ni	Pb	Zn
S1	70,289	8.81	0.165	13.0	67.6	27.8	37,737	591	31.9	20.4	86.6
S2	70,879	8.44	0.452	14.0	74.8	36.8	39,405	625	34.2	22.3	90.7
S3	72,134	8.65	1.79	14.5	73.5	32.3	40,070	712	35.0	33.2	83.1
S4	72,613	8.49	0.492	14.9	76.8	27.9	40,303	726	35.0	19.6	80.4
S5	62,044	6.41	0.220	12.1	58.2	21.1	33,428	597	27.5	17.0	64.1
S6	64,325	6.79	0.106	12.0	64.8	32.0	34,273	613	31.3	17.9	69.6
S7	77,529	7.77	0.044	14.0	75.1	28.4	40,084	389	38.1	20.5	74.4
S8	68,146	8.08	0.027	12.7	62.5	22.2	36,786	511	34.3	19.7	61.4
S9	72,666	9.40	0.044	14.1	74.2	36.6	40,838	785	40.1	22.9	74.9

The geoaccumulation index (I_{geo}) of Muller [97] has been calculated for the analyzed elements, by comparing current concentrations with pre-industrial levels, in order to estimate the metal contamination in sediments. The equation used for the calculation of I_{geo} is: $\log_2 (C_n/1.5\ B_n)$, where C_n is the measured content of element "n" and B_n the element's content in "average shale" [98]. Factor 1.5 is used because of possible variations in background values for a given element in the environment, as well as very small anthropogenic influences [99]. As shown in Figure 2, all sediments fall in class 0 for Al, Co, Cr, Cu, Fe, Mn, Ni, Pb, and Zn, therefore the area is not contaminated for these elements. Unlike the Hugli river, in other rivers of the Bengal Basin, such as Meghna and Brahmaputra, Cr exhibits higher I_{geo} values respect to the other elements [96]. For Cd, two stations fall in class 1 and three in class 2 exhibiting a moderate contamination for this element. In all stations, As falls in class 2 (moderate pollution). In this area, As contamination was already observed in previous studies and it is probably due to groundwater contamination [100]. This contamination can have natural origin, such as coal seams in Rajmahal basin and arsenic mineral in mineral rocks in the upper reaches of the Ganges river system. The highly reducing nature of groundwater would reduce As, causing the possible desorption of As [101].

Speciation Patterns

The potential environmental risk of trace elements in sediments is associated with both their total content and their speciation. The chemical partitioning of the considered elements (Al, As, Cd, Co, Cr, Cu, Fe, Mn, Ni, Pb, and Zn) from each extraction step has been described. Aluminum, Cr, and Fe are present mainly in the residual phase, representing 95.8–96.8%, 88.9–91%, and 83.0–94.7% of the total concentration, respectively, which implies that these elements are strongly linked to the inert fraction of the sediments. This result was in good agreement with data reported by several studies carried out worldwide in marine coastal areas [45, 46, 78, 102]. The high percentage of Fe in the residual phase indicates that most of the Fe exists as crystalline Fe peroxides (goethite, limonite, magnetite, etc.). The remaining Fe is associated with the reducible phase (mean, 11.25%). Large amounts of Fe accumulate in the residual phase probably because it is basically of natural origin (it is the most common element in the earth's crust).

Concentrations of Al, Fe, and Cr are very low in exchangeable phase (0.08%, 0.26%, and 1.72% as mean values, respectively), limiting their potential toxicity as pollutants. It should be noted that sediments always act as reservoir for elements; therefore, their potential risk of pollution to environment has always to be considered.

Arsenic, Co, Ni, and to some extent Zn, are found mainly in the residue (~50% of the total concentration). Nickel and Co are associated to the residue respectively for 56% and 74% of the total concentration, with a speciation similar in all the samples. A mean of 23% of Co is present in the phase 2. The highest percentage of labile Co (~13%) was measured in S_6 (Diamond Harbour) and S_8 (Gangasagar) and can be due to a recent input of this element. The dominant proportion of Ni in the residual phase is in agreement with the results of other studies [27, 46]. Nickel is present, apart from the residue, in phases 2 and 3 (about 10% in each phase). Arsenic is distributed mainly between the residual (mean 47%), the reducible, and the oxidizable phases (mean 19% and 22%, respectively). Acharyya et al. [101] observed that As is adsorbed to iron-hydroxide-coated sand grains and to clay minerals in the sediments of the Ganges delta from West Bengal. Among the studied elements, As is found with the greatest proportion in the oxidizable phase coinciding with organic and sulfur compounds. Arsenic is present in the phase 1 for about 10% of the total content, in station S_7 phase 1 percentage rises up to 16%. The lower land alluvial basin of the Ganges River is recognized as an arsenic-affected area. Arsenic in solution probably is easily entrapped in the fine grained organic-rich sediments deposited in the Ganges delta [101]. The percentage of silt (lower than 70% except in S_1 and S_2) may have contributed to a low retention of dissolved As since coarse sediments are less efficient at retaining As.

Cadmium was mainly present in the labile phase (more than 60%) in all the stations with the exception of station S_7, where the Cd labile percentage represents only 25% of the total concentration. Cadmium concentrations were negligible in phases 2 and 3. The highest labile Cd concentration was measured at station S_3, the closest to the city of Calcutta. Datta and Subramanian [96] found that the concentrations of elements in the non-detrital phases were higher in stations sampled in the Hugli river around Calcutta than in samples collected along Brahmaputra and Meghna rivers. The petroleum refinery, industrial, and mining effluents carried by the Hugli river may be responsible for this higher concentrations of non detrital fractions.

About 40% of the total Cu concentration is associated to the residue, while 33% of Cu is bound to Fe-Mn oxide and hydroxide (phase 2). The high percentage of Cu in the residue is likely due to the fact that Cu is easily chemisorbed on or incorporated in clay minerals [103]. All the samples showed lower Cu concentrations in exchangeable phase, with percentage ranging from 7% (S_7) to 22% (S_5), with a mean of 15%. Copper is characterized by high complex constant with organic matter thus it can be hypothesized that Cu is bound to labile organic matter such as lipids, proteins, and carbohydrates. On the other hand, high-element concentration in labile phase could be related to recent coastal input [39].

Manganese was found in all the four sediment phases, as observed by other researchers [45, 104]. Manganese is the most mobile element since it is present with the highest percentage (a mean of 42%) in the labile phase. This is probably because of the known close association of Mn with carbonates [105] as endorsed by other workers [69, 106]. In this phase, weakly sorbed Mn retained on sediment surface by relatively weak electrostatic interactions may be released by ion exchange processes and dissociation of Mn-carbonate phase [2]. The result indicates that considerable amount of Mn may be released into environment if conditions become more acidic [107]. The highest Mn labile percentage was measured in S_6 (57%). Differently, in S_7, Mn in the residue represents 65% of the total concentration, while the labile Mn is only 15%. A substantial Mn percentage was also found in the residue (mean 37.8%), followed by the reducible phase (14.7%), in which Mn exists as oxides and may be released if the sediment is subjected to more reducing conditions [108].

The major geochemical phase for Pb in these sediments was the Fe-Mn oxides phase (mean 55.7%) followed by the residual phase (mean 30.2%) while lower percentage of the total Pb are bound to exchangeable-labile (mean 5.3%) and oxidizable phases (mean 6.8%). At S_3 (Babughat), the reducible part is as high as 65% and only 19.9% of the total is associated with the residue. Atmospheric input as fallout from vehicular emission can be probably the major input of Pb for this station. The relatively high percentage of Pb in reducible phase is in agreement with the known ability of amorphous Fe–Mn oxides to scavenge Pb from solution [109, 110]. Caille et al. [111] observed that resuspension of anoxic sediment results in a rapid desorption of Pb and

Cu adsorbed to sulphides. Thus, a high element percentage in the reducible fraction is a hazard for the aquatic environment because Fe and Mn species can be reduced into the porewaters during early diagenesis by microbially mediated redox reactions [112]. Dissolution will also release Pb associated with oxide phases to the porewater possibly to the overlying water column [113] and to benthic biota [79]. The major sources of Pb are from intensive human activities, including agriculture in the drainage basin [114], auto exhaust emission together with atmospheric deposition [115]. In addition, a substantial contribution from the factories located in the upstream of the Hugli river dealing with Pb producing lead ingots and lead alloys play a vital role as referred by Sarkar et al. [116].

The percentage of Zn in residue is highly variable (38.5–70%) and the distribution pattern in each fraction showed the following order: residual>reducible>oxidizable> exchangeable and bound to carbonates. There was some difference in Zn speciation among the sampling sites: in stations S_1, S_2, and S_3 about 40% of Zn is present in the residue, while in the other stations this percentage increases to more than 60%. In station S_1, the exchangeable and oxidisable phases shared over 22% of the total Zn, whereas labile Zn was as low as 4.6% at S_7. A major part of Zn (16.3%) is associated with Fe–Mn oxide phase, because of the high stability constants of Zn oxides. Iron oxides adsorb considerable quantities of Zn and these oxides may also occlude Zn in the lattice structures [117].

The BCR procedure as discussed above showed satisfactory recoveries, detection limits, and standard deviations for determinations of heavy metals/ metalloid in the sediments. It is evident from the present results of the fractionation studies that the metals/metalloids in the sediments are bound to different fractions with different strengths leading to variations in mobility and availability and some of them show significant spatial variations subject to diverse environmental stresses. This type of association between metals and the sediments can be understood in detail by sequential extraction techniques. Hence the application of sequential extraction is fully justified as the quantification of different forms of metal is more meaningful than the estimation of its total metal concentrations. The strength values can, therefore, give a clear indication of sediment reactivity, which in turn assess the risk connected with the presence of metals in this wetland environment. The results obtained suggest the need for corrective remediation measures due to the higher accumulation of potentially dangerous metals/metalloids, which in most cases exceed the limits established by certain legislation.

Comparison with Sediment Quality Guidelines

Results obtained after total and sequential extraction are compared with Sediment Quality Guidelines (SQGs). Table 6 reports consensus-based values, such as TEC (concentration below which harmful effects on sediment-dwelling organisms were not expected) and PEC (concentration above which harmful effects on sediment-dwelling organisms were expected to occur frequently), and effect range-low and range-medium, such as ERL (concentrations below which adverse biological effects were observed in less than 10% of studies) and ERM (concentrations above which effects were more frequently observed in more than 75% of studies).

Table 6: Sediment Quality Guidelines concentrations with respect to total and labile element concentrations found in the analyzed samples (expressed as µg/g of dry weight)

Element	Phase	Si<TEC	TEC	Si<TEC< PEC	PEC	Si<ERL	ERL	ERL<Si< ERM	ERM
As	Total	All	9.79	None	33	S5,S6,S7,S8	8.2	S1,S2,S3, S4, S9	70
	Labile	All		None		All		None	
Cd	Total	S1,S2,S4,S5, S6,S7,S8,S9	0.99	S3	4.98	S1,S2,S4,S5, S6,S7,S8,S9	1.2	S3	9.6
	Labile	S1,S2,S4,S5, S6,S7,S8,S9		S3		S1,S2,S4,S5, S6,S7,S8,S9		S3	
Cr	Total	None	43.4	All	111	All	81	None	370
	Labile	All		None		All		None	
Cu	Total	S1,S4,S5,S7, S8	31.6	S2,S3,S6, S9	149	S1,S3,S4,S5, S6,S7,S8	34	S2,S9	270
	Labile	All		None		All		None	
Ni	Total	None	22.7	All	48.6	None	20.9	All	51.6
	Labile	All		None		All		None	
Pb	Total	All	35.8	None	128	All	46.7	None	218
	Labile	All		None		All		None	
Zn	Total	All	121	None	459	All	150	None	410
	Labile	All		None		All		None	

Comparing our results with the SQGs, it is revealed that for Pb and Zn in all the stations the measured concentrations are lower than both TEC and ERL. As regards Cd, concentration measured in station S_3 is higher than TEC and ERL but lower than PEC and ERM both in term of total and labile concentration. For this station, some possible toxic effect on benthic organism can be hypothesized, in particular because of the large amount of element bound to the most labile phase of the sediment. Considering Cu, some stations (S_2, S_3, S_6, and S_9) exhibit total concentrations higher than TEC but lower than PEC. Concentrations of Cu are higher than ERL but lower than ERM only in stations S_2 and S_9. Since only 7–22% of total Cu is bound to the labile phase, in all stations Cu labile concentrations are lower than TEC and ERL. Total As concentrations in stations S_1, S_2, S_3, S_4, and S_9 are higher than ERL value but lower than TEC value. Since more than 50% of total As is not found in the residue, attention should be paid to a change in the environment conditions which could induce a release of As from the sediments. Total Ni and Cr concentrations are higher than TEC (Ni is also higher than ERL) but lower than PEC (and ERM in the case of Ni) in all the stations. Nevertheless, more than 70% of Ni as well as 90% of Cr are present in the residual fractions, therefore adverse impacts on organisms is very much negligible.

Mean sediment quality guidelines quotients (mSQGQ) have been developed for assessing the potential effects of contaminant mixtures in sediments [118]: they are determined by calculating the arithmetic mean of the quotients derived by dividing the concentrations of chemicals in sediments by their respective SQGs. The probability of observing sediment toxicity can be estimated by comparing the mSQGQ in a sample to previously published probability tables. It is important to keep in mind that mSQGQs cannot be used to accurately predict the uptake and bioaccumulation of sediment-bound chemicals by fish, wildlife, and humans, even if there is considerable evidence that this assessment tool can be predictive of the presence or absence of toxic effects [118].

SQGQs are calculated for seven elements considering ERM as sediment quality guidelines (Table 7). The mean quotient values ranges from 0.16 in station S_5 to 0.24 in station S_3. Using PEC values instead of ERM, the mean SQGQ ranges from 0.25 in station S_5 to 0.38 in station S_3 (Table 7).

Table 7: Mean Sediment Quality Guidelines Quotients calculated for the nine stations using PEC and ERM as SQGs

Stations	SQG$_{OPE}$C	SQG$_{OER}$M
S1	0.30	0.19
S2	0.33	0.21
S3	0.38	0.24

S4	0.33	0.21
S5	0.25	0.16
S6	0.28	0.18
S7	0.32	0.21
S8	0.28	0.18
S9	0.34	0.22

Compiled data from multiple data sets reporting 10-day toxicity test conducted on amphipod species in saltwater showed that the incidence of toxicity for a range of SQGQ of 0.25–0.5 is ~35%, while for a mean SQGQ range from 0.1 to 0.25, the incidence of toxicity lowers to ~20%. Measures recorded in a survey of Biscayne Bay (port of Miami and the adjoining saltwater reaches of the lower Miami River, FL, USA) showed that the average amphipod survival (*Ampelisca abdita*) decreased slightly from the least contaminated (ERMQ <0.03) to the intermediate category, (ERMQ included in 0.03–0.2 range) then decreased greatly in the most contaminated sediments (ERMQ included in 0.2–2 range). Therefore, we can presume a low toxicity of sediments sampled in the nine stations for benthic organisms. It is important to note that the benthic response to contaminants covaried among stations with both the mean ERM quotients and the effect of natural factors, such as the sediment texture, TOC, and salinity [118].

Statistical Analyses

The relationships between variables and the differences between stations were evaluated by PCA. The analysis was performed on 36 objects (four sediment phases for nine stations) and 11 variables (Al, As, Cd, Co, Cr, Cu, Fe, Mn, Ni, Pb, Zn). Two significant components were identified explaining 68.3% and 14.5% of the total variance, respectively. By studying the loadings of the variables (Figure 2a) on the components it can be seen that all the elements except Cd, Mn, and Pb are significantly correlated.

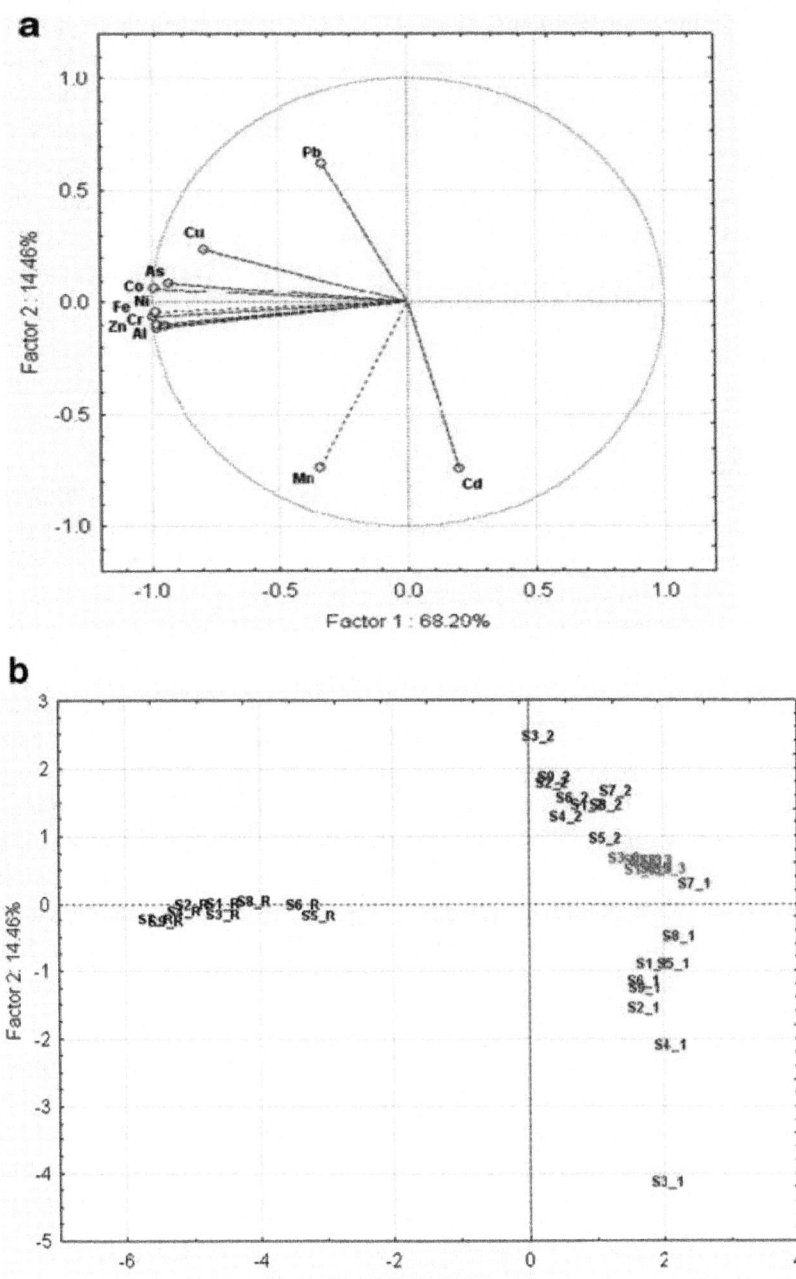

Figure 2: Principal component analysis: a) variable plot; b) score plot (phase 1, 2, 3 and 4, corresponding to labile, reducible, oxidizable and residue, are identifiable by 1, 2, 3 and 4 suffix and different colors in the score plot).

Unlike the other elements, most of Cd and Mn is present in the first phase: labile Cd and Mn represent more than 60% and 40% of the total concentration, respectively, except in station S_7. Cadmium and Mn speciation can be ascribed to their considerable affinity for carbonates. Lead is the only element which is bound to the reducible phase for more than 50%. Lead is a very reactive element in water column and, having scavenging type behavior, is easily bound to hydroxy- and oxyligands. Copper is positively and significantly correlated with all elements except Cd and Mn, but with lower correlation coefficients (0.66–0.81).

In the score plot (Figure 2b) phases 1, 2, 3, and 4 (corresponding to labile, reducible, oxidizable, and residual phases respectively) are identifiable by 1, 2, 3, and 4 suffix, respectively. In all stations, residue concentrations were characterized by negative values of PC1 and consequently by high concentrations of Al, As, Co, Cr, Fe, Ni, and Zn. Conversely, in the positive PC1 semi-axis labile and oxidizable metal concentrations, which represent a small percentage of the total elements, are distributed. For all stations, reducible concentrations are distributed along the positive PC2 semi-axis, i.e., high Pb concentrations, with a maximum for station S_3 and a minimum for S_5. The group formed by elements bound to organic matter and sulphides (phase 3) is characterized by low values of both PC1 and PC2. Therefore, a low percentage of elements (higher than 20% exclusively for all As data and for Zn in stations S_1 and S_3) is bound to the oxidizable phase, suggesting the presence of an oxidant environment. High Mn and Cd concentrations are associated with negative values of PC2, therefore a relatively high concentration of labile Mn and Cd is present in all samples (in particular in S_3), except station S_7. Samples are prevalently grouped in relation to the sediment geochemical phase, suggesting a similar element speciation among the stations. Station S_7 represents an exception, in fact the labile fraction is closely associated to the oxidizable phase group.

A HCA was carried out by applying Euclidean distances to quantitatively identify specific groups of similar stations. In the dendrogram of the sampling stations (Figure 3), we can note two main clusters: the first represented by station S_7, characterized by the highest element percentage bound to residue, and the second constituted by all the remainder stations. In the second group, a subgroup formed by station S_5 and S_6 can be individuated.

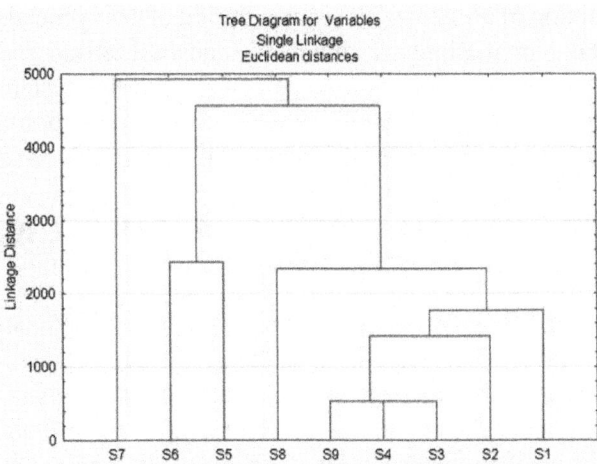

Figure 3: Dendrogram indicating linkage of sites on the basis of element concentrations.

Station S_7 was sampled in a marine coastal environment; it is characterized by a peculiar grain size percentage respect to the other stations, being the sand percentage as high as 98.6%. In general, the concentrations of elements are much higher in fine than in coarse fraction because the fine fraction larger specific surface facilitates absorption processes. As previously noted by Ramirez et al. [39], this pattern is particularly evident for Cd. It is interesting to note that the marine coastal stations S_8 and S_9 are more similar to river stations than to station S_7. Both the stations S_8 and S_9 are, in fact, located in front of the Hugli river runoff, while station S_7 is located easternmost and probably is less influenced by the Hugli river discharge.

Conclusion and Recommendation

The study provides valuable information on the potential mobility of trace elements in sediments collected along the stretch of Hugli River and in the Sundarban mangrove wetland (northeastern part of the Bay of Bengal). The results obtained adopting BCR sequential extraction method provided the following important information: (i) Al, Cr, and Fe were found mostly in the residual phase while the other elements were found in the four phases sharing different proportions; (ii) the dominant Cd, Mn, and Pb proportion was found in the non-residual fractions and (iii) Mn had the highest percentage in the labile phase. This is worthwhile to mention that coastal environment of West Bengal is considerably constrained due to implementation of dredging, construction of port/ harbor and other industrial activities. The authors strongly recommend for periodical monitoring on the bioavailability and mobility of trace elements, control the mixing of effluent of the concentration of heavy metals in the region, environmental remediation, treatment of industrial effluent and municipal wastewater for effective management of this estuarine system. It is wisely suggested that an environmental recovery framework should be urgently implemented to avoid extension of heavy metal contamination (especially As).

CASE STUDIES FROM OTHER COASTAL REGIONS IN INDIA

Although the importance of metal speciation and fractionation has been realized in developed countries, the subject has not really taken off in India and only few references are available on the speciation of metals in Indian rivers. Speciation of selected heavy metals geochemistry in surface sediments (n=10 was studied by Venkatramanan et. al. [119] from Tirumalairajan river estuary, east coast of India. The results obtained from sequential extraction showed that a larger portion of the metals were associated with the residual phase, although they are available in other fractions too.

Trace metal fractionation in the Pichavaram mangrove–estuarine sediments in southeast coast of India was studied by Ranjan et. al. [120] considering the pronounced changes due to occurrence of tsunami (2004). A 5-step sequential extraction procedure was applied to assess the effects of tsunami on mobility and redistribution of selected elements (Cd, Cr, Cu, Fe, Mn, Ni, Pb, and Zn) in coastal sediments revealed that metals in the residual fraction (lattice bound) had the highest concentration suggesting their non-availability and limited biological uptake in the system. Majority of the metals (except Mn) do not constitute a risk based on the different geochemical indices.

Fractionation of selected metals in the sediments of Cochin estuary and Periyar River (southwest coast of India) was studied by Mohan et. al. [121]. The results reveal that remobilization potential of metals bound is in the range of low to medium risk to various sedimentary phases is different and is based on bond strength. Therefore, the strength values can give a clear indication of sediment reactivity that can be used to assess the risk related with metals to the aquatic organisms.

RISK ASSESSMENT CODE (RAC)

The risk assessment code (RAC) mainly applies the sum of the exchangeable and carbonate bound fractions for assessing the availability of metals in sediments. These fractions are considered to be weakly bonded metals which may equilibrate with the aqueous phase and thus become more rapidly bioavailable [11, 33]. This is important because the fractions introduced by anthropogenic activities, such as agricultural runoff and tourism, are typified by the adsorptive, exchangeable, and bound to carbonate fractions, which are weakly bonded metals that could equilibrate with the aqueous phase and thus become more rapidly bioavailable [122]. According to RAC guideline (Table 8), for any metal, soil/sediment which can release in exchangeable and carbonate fractions, less than 1% of the total metal will be considered safe for the environment and soil/sediment with 11-30% carbonate and exchangeable fractions will be at medium risk to the environment. On the contrary, soil/sediment releasing in the above fractions more than 50% of the total metal has considered being highly dangerous, which can be easily enter the food chain [123].

Table 8: Criteria of Risk Assessment Code [123]

Grade	Exchangeable and bounded to carbonate metal (%)	Risk
I	<1	No risk
II	1 – 10	Low risk
III	11 – 30	Medium risk
IV	31 – 50	High risk
V	>50	Very high risk

Heavy-Metal Fractionation in surface sediments was studied by Dhanakumar et. al. [124] in the Cauvery river estuarine region, southeastern coast of India. The results revealed that most of the samples fall under the category from low- to high-risk class and from low-risk to very high-risk class in terms of labile fractions of Pb as well as Zn and Cu, respectively.

CONCLUSION

From the above discussion it is revealed that geochemical fractionation approach to the chemical speciation has provided a useful tool and opens a new dimension in assessing the potential mobility/bioavailability of heavy metals and metalloids in soils/sediments and opens a new dimension in the field of ecology and environmental chemistry. More efficient, non-laborious and time saving processes techniques in this field of chemical speciation are also coming up to get valid information regarding geochemical behavior of soils/sediments. Besides geochemical fractionation, Dezileau et al. [125] opined that total Fe or Fe/Al may be used to infer millennial-scales climate changes in the south eastern pacific while performing sequential extraction of Fe in marine sediments from the Chileau continental margin. However, the chemical partitioning should be carefully used in the assessment of environmental pollution as large amount of metals may naturally occur as anthropogenic fractions (including loosely bonded ions, sulfide ions and metals associated with sediments).

ACKNOWLEDGEMENTS

The authors gratefully acknowledge full support and cooperation of the Springer press, UK for extending permission in publishing the research paper of the journal Environmental Monitoring & Assessment, vol. 184(12), pp:7561-77, 2012.

This study was partially supported by the European Fund for Economic and Regional Development (FEDER) through the Program Operational Factors of Competitiveness (COMPETE) and National Funds through the Portuguese Foundation for Science and Technology (PEST-C/MAR/UI 0284/2011, FCOMP 01 0124 FEDER 022689).

REFERENCES

1. Gleyzes C, Tellier S, Astruc M. Fractionation studies of trace elements in contaminated soils and sediments: a review of sequential extraction procedures. Trends in Analytical Chemistry 2002; 21 (6) 451-467.

2. Tessier A, Campbell PGC, Bisson M. Sequential extraction procedure for the speciation of particulate trace metals. Analytical Chemistry 1979; 51 844-851.

3. Sahuquillo A, Rigol A, Rauret G. Overview of the use of leaching/ extraction tests for risk assessment of trace metals in contaminated soils and sediments. Trends in Analytical Chemistry 2003; 22(3) 152-159.

4. Calmano W, Hong J, Förstner U. Binding and mobilization of heavy metals in contaminated sediments affected by pH and redox potential. Water Science and Technology 1993; 28(8/9) 53-58.

5. Eggleton J, Thomas KV. A review of factors affecting the release and bioavailability of contaminants during sediment disturbance events. Environment International 2004; 30 973-980.

6. Ure AM, Quevauviller P, Muntau H, Griepink B. Speciation of heavy metals in solids and harmonization of extraction techniques undertaken under the auspices of the BCR of the Commission of the European Communities. International Journal of Environmental Analytical Chemistry 1993; 51 135-151.

7. Florence TM. The speciation of trace elements in waters. Talanta 1982;

29 345-69.

8. Lung W. Speciation analysis – why and how? Fresenius' Journal of Analytical Chemistry 1990; 337 557-564.

9. Schroeder WH. Development in the speciation of mercury in natural waters. Trends in Analytical Chemistry 1989; 8 339-342.

10. Rauret G, Rubio R, Lopez-Sanchez JF, Cassassas E. Determination and speciation of copper and lead in sediments of a Mediterranean River (River Tenes, Catalonia, Spain). Water Research 1988; 22(4) 449-51.

11. Pardo R, Barrado E, Perez L, Vega M. Determination and association of heavy metals in sediments of the Pisuerga River. Water Research 1990; 24(3) 373-379.

12. Jardo CP, Nickless G. Chemical association of Zn, Cd, Pb and Cu in soils and sediments determined by the sequential extraction technique. Environmental science & Technology Letters 1989; 10 743-52.

13. Tessier A, Campbell PGC, Bisson M. Heavy metal speciation in the Yamaska and St. Francois rivers (Quebec). Canadian Journal of Earth Sciences 1980; 17 90-105.

14. Elsokkary LH, Muller G. Assessment and speciation of chromium, nickel, lead and chromium in the sediments of the river Nile, Egypt. Science of the Total Environment 1990; 97/98 455-63.

15. Ure AM. Single extraction schemes for soil analysis and related applications. Science of the Total Environment 1996; 178(1/3) 3-10.

16. Rauret G. Extraction procedures for the determination of heavy metals in contaminated soil and sediment. Talanta 1998; 46(3) 449-455.

17. Quevauviller PH, Lachica M, Barahona E, Rauret G, Ure A, Gomez A, Muntau H. Interlaboratory comparision of EDTA and DTPA procedures prior to certification of extractable trace elements in calcareous soil. Science of the Total Environment 1996; 178(1/3) 127-132.

18. Quevauviller PH, van der Slootr HA, Ure A, Muntau H, Gomez A, Rauret G. Conclusions of the workshop: harmonization of leaching/ extraction tests for environmental risk assessment. Science of the Total Environment 1996; 178(1/3) 133-139.

19. Bacon JR, Davidson CM. Is there a future for Sequential chemical extraction? Analyst 2008; 133 25-46

20. Förstner U, Lechsber R, Davis RA, Hermitte PL. (eds.), Chemical methods for assessing bioavailable Metals in Sludges. Elsevier, London, 1985.

21. Meguellati M, Robbe D, Marchandise P, Astruc M, Proceedings

International Conference on Heavy Metals in the Environment, Heidelberg CEP Consultants, Edinburgh, 1983, p. 1090.

22. Filgueiras AV, Lavilla I, Bendicho C. Chemextraction for metal partitioning in environmental solid samples, Journal of Environmental Monitoring 2002; 4 832-857.

23. Lopez-sanchez JF, Rubio R, Rauret G. Comparison of two sequential extraction procedures for trace metal partitioning in sediments. International Journal of Environmental Analytical Chemistry 1993; 51 113-121.

24. Hlavay J, Polyak K. Chemical speciation of elements in sediment samples collected at Lake Balaton. Microchemical Journal 1998; 58 281-290.

25. Tokalioglu S, Kartal S, Elçi L. Determination of heavy metals and their speciation in lake sediments by flame atomic absorption spectrometry after a four-stage sequential extraction procedure, Analytical Chimica Acta 2000; 413 33-40.

26. Belazi AU, Davidson CM, Keating GE, Littlejohn D. Determination and speciation of heavy metals in sediments from the Cumbrian coast, NW England, UK. Journal of Analytical Atomic Spectrometry 1995; 10 233-240.

27. Mester Z, Cremisini C, Ghiara E, Morabito R. Comparison od two sequential extraction procedures for metal fractionation in sediment samples. Analytical Chimica Acta 1998; 259 133-142.

28. Gomez-Ariza JL, Giraldez I, Sanchez-Rodas D, Morales E. Comparison of the feasibility of three extraction procedures for trace metal partitioning in sediments from southwest Spain, Science of Total Environment 2000; 246 271-283.

29. Zhang T, Shan X, Li F. Comparison of two sequential extraction procedures for speciation analysis of metals in soils and plant availability. Communications in Soil Science and Plant Analysis 1998; 29 1023-1034.

30. Albores AF, Cid BP, Gomez P, Lopez EF. Comparison between sequential extraction procedures and single extrations for metal partitioning in sewage sludge samples. Analyst 2000; 125 1353-1357.

31. Ho MD, Evans GJ. Operational speciation of cadmium, copper, lead and zinc in the NIST standard reference materials 2710 nad 2711 (Montana soil) by the BCR sequential extraction procedure and flame atomic spectrometry. Analytical Communications 1997; 34 353-364.

32. Zemberyova M, Bartekov J, Hagarov I. The utilization of modified BCR three-step sequential extration procedure for the fractionation of Cd, Cr,

Cu, Ni, Pb and Zn in soil reference materials of different origins, Talanta 2006; 70 973-978.

33. Singh KP, Mohan D, Singh VK, Malik A. Studies on distribution and fractionation of heavy metals in Gomti river sediments- a tributary of the Ganges India. Journal of Hydrology 2005; 312 14-27.

34. Davidson CM, Duncan AL, Littlejohn D, Ure AM, Garden LM. A critical evaluation of the three-stage BCR sequential extraction procedure to assess the potential mobility and toxicity of heavy metals in industrially-contaminated land. Analytica Chimica Acta 1998; 363(1) 45-55.

35. McLaren RG, Crawford D. Studies on soil copper I. The fractionation of copper in soils. Journal of Soil Science 1973; 24(2), 172-181.

36. Kersten M, Förstner U. Speciation of trace elements in of water-soluble organic components. sediments and combustion waste.In Ure AM, Davidson CM (ed.) Chemical speciation in the environment. Blackie Academic and Professional, Glasgow, UK. 1995; 234-275.

37. Ianni C, Magi E, Rivaro P, Ruggieri N. Trace metals in Adriatic coastal sediments: distribution and speciation pattern. Toxicological and Environmental Chemistry 2000; 78 73-92.

38. Ianni C, Ruggieri N, Rivaro P, Frache R. Evaluation and comparison of two selective extraction procedures for heavy metal speciation in sediments. Analytical Sciences 2001; 17 1273-1278.

39. Ramirez M, Massolo S, Frache R, Correa J. Metal speciation and environmental impact on sandy beaches due to El Salvador copper mine, Chile. Marine Pollution Bulletin 2005; 50 62-72.

40. Borovec Z, Tolar V, Mraz L. Distribution of some metals in sediments of the central part of the Labe (Elbe) River, Czech Republic. Ambio 1993; 22 200-205.

41. Campanella L, D'Orazio D, Petronio BM, Pietrantonio E. Proposal for a metal speciation study in sediments. Analytica Chimica Acta 1995; 309 387-393.

42. Zdenek B. Evaluation of the concentrations of trace elements in stream sediments by factor and cluster analysis and the sequential extraction procedure. Science of the Total Environment 1996; 177 237-250.

43. Gomez-Ariza JL, Giraldez I, Sanchez-Rodas D, Moralesm E. Metal sequential extraction procedure optimized for heavily polluted and iron Oxide rich sediments. Analytica Chimica Acta 2000; 414 151-164.

44. Lopez-Sanchez JF, Sahuquillo A, Fiedler HD, Rubio R, Rauret G, Muntau H, Quevauviller P. CRM 601, a stable material for its extractable content

of heavy metals. Analyst 1998; 123 1675-1677.

45. Usero J, Gamero M, Morillo J, Gracia I. Comparative study of three sequential extraction procedures for metals in marine sediments. Environment International 1998; 24 478-496.

46. Martin R, Sanchez DM, Gutierrez AM. Sequential extraction of U, Th, Ce, La and some heavy metals in sediments from Ortigas River, Spain. Talanta 1998; 46 1115-1121.

47. Agnieszka S, Wieslaw Z. Application of sequential extraction and the ICPAES method for study of the partitioning of metals in fly ashes. Microchemical Journal 2002; 72 9-16.

48. Templeton DM, Ariese F, Cornels R. IUPAC guidelines for terms related to chemical Speciation and Fractionation of elements. Pure and Applied Chemistry 2001; 72 1453-1470.

49. Bordas F, Bourg ACM. A critical evaluation of sample for storage of Contaminated Sediments to Be investigated for the potential mobility of their heavy metal load. Water. Air. Soil. Pollution 1998; 103 137-149.

50. Kersten M, Frostner U. Chemical fractionation of heavy metals in anoxic estuarine and coastal sediments. Water Science and Technology 1986; 18 121-130.

51. Dold B. Speciation of the most soluble phases in a sequential extraction procedure adapted for geochemical studies of copper sulfide mine waste. Journal of Geochemical Exploration 2003; 80 55-68.

52. Favas PJC, Pratas J, Gomes MEP, Cala V. Selective chemical extraction of heavy metals in tailings and soils contaminated by mining activity: Environmental implications. Journal of Geochemical Exploration 2011; 111 160-171.

53. Pardo R, Vega M, Debán L, Cazurro C, Carretero C. Modelling of chemical fractionation patterns of metals in soils by two-way and three-way principal component analysis. Analytica Chimica Acta 2008; 606 26-36.

54. Cuong TD, Obbard JP. Metal speciation in coastal marine sediments from Singapore from Singapore using a modified BCR-sequential extraction procedure. Applied Geochemistry 2006; 21 1335-1346.

55. Dapaah RK, Takano N, Ayame A. Solvent extraction of Pb (II) from acid medium with zinc Hexamethylenedithiocarbamate followed by back-extraction and subsequent determination by FAAS. Analytica Chimica Acta 1999; 386 281-286.

56. Gomez-Ariza JL, Giraldez I, Sanchez-Rhodes D, Morales E. Metal

readsorption and re-distribution during analytical fractionation of trace elements in toxic estuarine sediments. Analytica Chimica Acta 1999; 399 295-307.

57. Cheam V, Lechner J, Sekerka I, Desrosiers R, Nriagu J. Development of laser- excited atomic fluorescence spectrometer and a method for the direct determination of lead in Great Lake waters. Analytica Chimica Acta 1992; 269 129-136.

58. Fischer E, Van D, Berg CMG. Anodic Stripping Voltammetry of Pb and Cd using a Hg film electrode and thiocyanate. Analytica Chimica Acta 1999; 385 273-280.

59. Morales MM, Mart P, Llopis A, Compos L, Sagrado S. An environmental study by factor Analysis of surface sea waters in the Gulf of Valencia (Western Mediterranean). Analytica Chimica Acta 1999; 394 109-117.

60. Hirade M, Chen Z, Sugimoto K, Kawaguchi H. Co precipitation with tin (IV) hydroxide followed by removal of tin carrier for the Determination of trace heavy metals by graphite-furnace atomic absorption Spectrometry. Analytica Chimica Acta 1980; 302 103-107.

61. Ridout PS, Jones HR, Williams JG. Determination of trace elements in a marine reference material of lobster hepatopancreas (TORT-1) using inductively coupled plasma mass spectrometry. Analyst 1988; 113 1383-1386.

62. Sakao SY, OgawaY, Uchida H. Determination of trace elements in seaweed samples by inductively coupled plasma mass spectrometry. Analytica Chimica Acta 1999; 355 121-127.

63. Baeyens W, Monteny F, Leermakers M, Bouillon S. Evalution of sequential extractions on dry and wet sediments. Analytical and Bioanalytical Chemistry 2003; 376 890- 901.

64. Guevara-Riba A, Sahuquillo A, Rubio R, Rauret G. Assessment of metal mobility in dredged harbour sediments from Barcelona, Spain. Science of the Total Environment 2004; 321 241-255.

65. Idris AM, Eltayeb MAH, Potgieter-Vermaak SS, Grieken R, Potgieter JH. Assessment of heavy metal pollution in Sudanese harbours along the Red Sea coast. Microchemical. Journal 2007; 87 104-112.

66. Wepener V, Vermeulen LA. A note on the concentrations and bioavailability of selected metals in sediments of Richards Bay Harbour, South Africa. Water SA 2005; 31 589-595.

67. Esslemont G. Heavy metals in seawater, marine sediments and corals from the Townsville section, Great Barrier Reef Marine Park, Queensland.

Marine Chemistry 2000; 71 215- 231.

68. Coetzee PP. Determination and speciation of heavy metals in sediments of the Hartebeespoort Dam By sequential extraction. Water SA 1993; 19 291-300.

69. Salmons W, Förstner U. Trace metal analysis on polluted sediments. Part II:evaluation of Environmental impact. Environmental science and Technology letters 1980; 1 14-24.

70. Li X, Thornton I. Chemical partitioning of trace and major elements in soils contaminated by mining and smelting activities. Applied geochemistry 2000; 16 1693-1706.

71. Kiratli N, Ergin M. Partitioning of heavy metals in surface Black Sea sediments. Applied Geochemistry 1996; 11 775-788.

72. Rauret G, Lopez-Sanchez JF. New sediment and soil CRMs for extractable Trace metal content. International Journal of Environmental Analytical Chemistry 2001; 79 81-95.

73. Salmons W. Adoption of common schemes for single and sequential extractions of Trace metals in soil and sediments. International Journal of Environmental Analytical Chemistry 1993; 51 3-4.

74. Fiedler HD, Lopez-Sanchez JF, Rubio R, Rauret G, Quevauviller PH. Study of the stability of extractable trace metal contents in a river sediment using Sequential extraction. Analyst 1994; 119 1109-1114.

75. Ramos L, Hernandez LM, Gonzalez MJ. Sequential fraction of copper, lead, copper, Cadmium and zinc in soils from or near Donana National Park. Journal of Environmental Quality 1994; 23 7-50.

76. Tu Q, Shan XZ, Ni Z. Evaluation of a sequential extraction procedure for the Fractionationation of amorphous iron and manganese oxides and organic matter in soils. The Science of The total Environment 1994; 151 159-165.

77. Wang S, Jia Y, Wang S, Wang X, Wang H. Fractionation of heavy metals in shallow marine sediments from Jinzhou Bay, China. Journal of Environmental Science (China) 2010 22 23-31.

78. Yuan CG, Shi JB, He B, Liu JF, Liang LN. Speciation of heavy metals in marine sediments from the East China Sea by ICP-MS with sequential extraction. Environment International 2004; 30 769-783.

79. Jones B, Turki A. Distribution and Speciation of heavy metals in surficial sediments from the Tees Estuary, North – East England. Marine Pollution Bulletin 1997; 34 768-779.

80. Ajay SO, Van Loon GW. Studies on redistribution during the analytical

fractionation of metals in sediments. The Science of the Total Environment 1989; 87 171-187.

81. Pempkowiak J, Sikora A, Biernacka E. Speciation of heavy metals in marine sediments vs their bioaccumulation by mussels. Chemosphere 1999; 39 313- 321.

82. Forstner U, Wittmann GTW. Metal Pollution in the Aquatic Environment, Springer- Verlag, Berlin. Springer-Verlag, Heidelberg; 1981.

83. Fedotov PS, Zavarzina, a AG, Spivakov BYa, Wennrich, b R, Mattusch J, De K, Titzeb PC, Demin VV. Accelerated fractionation of heavy metals in contaminated soils and sediments using rotating coiled columns, Journal of Environmental Monitoring 2002; 4 318-324.

84. Nemati K, Kartini N, Bakar A, Abas MR, Sobhanzadeh E. Speciation of heavy metals by modified BCR sequential extraction procedure in different depths of sediments from Sungai Buloh, Selangor, Malaysia, Journal of Hazardous Materials 2011; 192(1) 402-410.

85. Mossop KF, Davidson CM. Comparison of original and modified BCR sequential extraction procedures for the fractionation of copper, iron, lead, manganese and zinc in soils and sediments. Analytica Chimica Acta 2003; 478 (1) 111-118.

86. Walkey A, Black TA. An examination of the Dugtijaraff method for determining soil organic matter and proposed modification of the chronic and titration method. Soil Science 1934; 37 23-38.

87. Farnham IM, Johannesson KH, Singh AK, Hodge VF, Stetzenbach KJ. Factor analytical approaches for evaluating groundwater trace element chemistry data. Analytica Chimica Acta 2003; 490 123-138.

88. Ragno G, De Luca M, Ioele G. An application of cluster analysis and multivariate cassification methods to spring water monitoring data. Microchemical Journal 2007; 87 119-127.

89. Jonathan MP, Ram Mohan V. Heavy metals in sediments of the inner shelf off the Gulf of Mannar, Southeast coast of India. Marine Pollution Bulletin 2003; 46 263-268.

90. Sunil Kumar R. Distribution of organic carbon in the sediments of Cochin mangroves, south west coast of India. Indian Journal of Marine Science 1996; 25 274-276.

91. Janaki-Raman D, Jonathan MP, Srinivasalu S, Armstrong-Altrin J S, Mohan SP, Ram-Mohan V. Trace metal enrichments in core sediments in Muthupet mangroves, SE coast of India: Application of acid leachable technique. Environmental Pollution 2007; 145 245-257.

92. Sarkar SK, Bilinski SF, Bhattacharya A, Saha M, Bilinski H. Levels of elements in the surficial estuarine sediments of the Hugli river, northeast India and their environmental implications. Environment International 2004; 30 1089-1098.

93. Tam NFY, Wrong YS. Spatial variation of heavy metalsin surfe sediments of Hong, Kong mangrove swamps. Environmental Pollution 2002; 110 195-205.

94. Subramanian V, Mohanachandran G. Heavy metals distribution and enrichment in the sediments of southern east coast of India. Marine Pollution Bulletin 1990; 21 324-330.

95. Chatterjee M, Massolo S, Sarkar SK, Bhattacharya AK, Bhattacharya BD, Satpathy KK, Saha S. An assessment of trace element contamination in intertidal sediment cores of Sunderban mangrove wetland, India for evaluating sediment quality guidelines. Environmental Monitoring and Assessment 2009; 150 307-322.

96. Datta DK, Subramanian V. Distribution and fractionation of heavy metals in the surface sediments of the Ganges–Brahmaputra–Meghna river system in the Bengal Basin. Environmental Geology 1998; 36 93-101.

97. Müller G. Schwermetalle in den sedimenten des Rheins-Veranderungen seit. Umschan Verlag 1979; 79 133-149.

98. Salomon W, Förstner U. Metals in the hydrocycle. Berlin: Springer; 1984.

99. Buccolieri A, Buccolieri G, Cardellicchio N, Dell'atti A, Leo AD, Maci A. Heavy metals in marine sediments of Taranto Gulf (Ionian Sea, Southern Italy). Marine Chemistry 2006; 99 227-235.

100. Dowling CB, Poreda RJ, Basu AR, Aggarwal PK. Geochemical study of arsenic release mechanisms in the Bengal Basin groundwater. Water Resources Research 2002; 38(9) 1173-1190.

101. Acharyya SK, Lahiri S, Raymahashay BC, Bhowmilk A. Arsenic toxicity of groundwater in parts of the Bengal basin in India and Bangladesh: the role of Quaternary stratigraphy and Holocene sea-level fluctuation. Environmental Geology 2000; 39 231-238.

102. Takarina ND, Browne DR, Risk MJ. Speciation of heavy metals in coastal sediments of Semarang, Indonesia. Marine Pollution Bulletin 2004; 49 854-874.

103. Pickering WF. Metal ion speciation—soil and sediments (a review). Ore Geology Reviews 1986; 1 83-146.

104. Ngiam LS, Lim PE. Speciation patterns of heavy metals in tropical estuarine anoxic and oxidized sediments by different sequential extraction

schemes. Science of the Total Environment, 2001; 275 53-61.

105. Dassenakis M, Adrianos H, Depiazi G, Konstantas A, Karabela M, Sakellari A, Scoullos M. The use of various methods for the study of metal pollution in marine sediments, the case of Euvoikos Gulf, Greece. Applied Geochemistry 2003; 18 781-794.

106. Morillo J, Usero J, Gracia I. Heavy metal distri-bution in marine sediments from the southwest coast of Spain. Chemosphere 2004; 55 431-442.

107. Thomas RP, Ure AM, Davidson CM, Littlejoh D, Rauret G, Rubio R, López-Sánchez JF. Three-stage sequential extraction procedure for the deter-mination of metals in river sediments. Analytica Chimica Acta 1994; 286 423-429.

108. Panda D, Subramanian V, Panigrahy RC. Geochemical fractionation of heavy metals in Chilka Lake (east coast of India) – a tropical coastal lagoon. Environmental Geology 1995; 26 199-210.

109. Dawson EJ, Macklin MG. Speciation of heavy metals in floodplain and flood sediments: a reconnaissance survey of the Aire Valley, West Yorkshire, Great Britain Environmental Geochemistry and Health 1998; 20 67-76.

110. Ramos L, González M, Hernández L. Sequential extraction of copper, lead, cadmium, and zinc in sediments from Ebro River (Spain): relationship with levels detected in earthworms. Bulletin of Environmental Contamination and Toxicology 1999; 62 301-308.

111. Caille N, Tiffreau C, Leyval C, Morel JL. Solubility of metals in an anoxic sediment during prolonged aeration. Science of the Total Environment 2003; 301 239-250.

112. Canfield DE. Reactive iron in marine sediments. Geochimica et Cosmochimica Acta 1989; 53 619-632.

113. Petersen W, Wallman K, Li PL, Schroeder F, Knauth HD. Exchange of trace elements at the sediment– water interface during early diagenesis processes. Marine and Freshwater Research 1995; 46 19-26.

114. Monbet P. Mass balance of lead through a small macrotidal estuary: the Morlaix River estuary (Brittany, France). Marine Chemistry 2006; 98 59-80.

115. Adriano DC. Trace elements in terrestrial environments. New York: Springer; 1986.

116. Sarkar SK, Saha M, Takada H, Bhattacharya A, Mishra P, Bhattacharya B. Water quality management in the lower stretch of the river Ganges, east coast of India: an approach through environmental education. Journal of

Cleaner Production 2007; 15 1559-1567.

117. Banerjee ADK. Heavy metal levels and solid phase speciation in street dusts of Delhi, India. Environmental Pollution 2003; 123(1) 95-105.

118. Long ER, Ingersoll CG, MacDonald DD. Calculation and uses of mean sediment quality guideline quotients: a critical review. Environmental Science and Technology 2006; 40 1726-1736.

119. Venkatramanan S, Ramkumar T, Anithamary I, Jonathan MP. Speciation of selected heavy metals geochemistry in surface sediments from Tirumalairajan river estuary, east coast of India. Environmental Monitoring and Assessment 2013; 185(8) 6563-6578.

120. Ranjan RK, Singh G, Routh J, Ramanathan AL. Trace metal fractionation in the Pichavaram mangrove–estuarine sediments in southeast India after the tsunami of 2004. Environmental Monitoring and Assessment 2013 (article in press).

121. Mohan M, Augustine T, Jayasooryan KK, Chandran MSS, Ramasamy EV. Fractionation of selected metals in the sediments of Cochin estuary and Periyar River, southwest coast of India, Environmentalist 2012; 32 383-393.

122. Hseu ZY. Extractability and bioavailability of zinc over time in three tropical soils incubated with biosolids. Chemosphere 2006; 63 762-771.

123. Perin G, Craboledda L, Lucchese M, Cirillo R, Dotta L, Zanette ML, Orio AA. Heavy metal speciation in the sediments of Northern Adriatic Sea- a new approach for environmental toxicity determination, in: T.D. Lekkas (Ed.), Heavy Metal in the Environment 1985; 2 454-456.

124. Dhanakumar S, Murthy KR, Solaraj G, Mohanraj R. Heavy-Metal Fractionation in Surface Sediments of the Cauvery River Estuarine Region, Southeastern Coast of India, Arch Environmental Contamination Toxicology 2013; 65 14-23.

125. Dezileau L, Pizarro C Rubio MA. Sequential extraction f iron in marine sediments from the Chilen continental margin. Marine Geology 2007; 241 111-116.

Chapter 2

CHEMICAL AND HYDRAULIC BEHAVIOR OF A TROPICAL SOIL COMPACTED SUBMITTED TO THE FLOW OF GASOLINE HYDROCARBONS

Wisley Moreira Farias[1], Geraldo Resende Boaventura[1], Éder de Souza Martins[2], Fabrício Bueno da Fonseca Cardoso[3], José Camapum de Carvalho[1] and Edi Mendes Guimarães[1]

[1]Universidade de Brasília, Brasília-DF, Brazil

[2]Embrapa/ Cerrados, Planaltina –DF, Brazil

[3]ANA-Agência Nacional de Águas –Brasília-DF, Brazil

INTRODUCTION

Gasoline is a fuel comprised basically of hydrocarbons such as aromatic, olefinic and saturated compounds of a carbon chain comprised of 4 to 12 atoms. The aromatic compounds such as benzene, toluene, ethylbenzene, o-, m-, p-xylene (BTEX) are harmful to human health (Cairney et. al., 2002). As these compounds are harmful to health, the legislation becomes restrictive. The U.S. Environmental Protection Agency for drinking water (US EPA) establishes the maximum concentration of benzene in 5µg.L^{-1}. In Brazil the Ordinance of the Ministry of health number 2,914 in 12th December 2011, stipulates that the maximum allowable concentration of benzene is 5 µg.L^{-1} regulation of drinking water contaminant. Soil in contaminated residential areas, Brazil has been adopting as intervention guide value, the concentration of benzene 0.08 mg.kg^{-1} set by the State of São Paulo in 2001. This value indicates the intervention limit of contamination where there is potential risk to human health.

Brazil produces type-C gasoline which is different than other types due to its anhydrous alcohol content (ethanol), in the proportion of 25% (Farias, 2003). The alcohols are soluble in water, and have a significant mobility potential to percolate through the soil until reaching underground water (Ulrich,

1999; Corseuil and Fernandes, 1999). The alcohol in gasoline in an aqueous medium promotes co-solvency which is the increase in the solubility of the hydrocarbons in the gasoline in an aqueous solution (Banerjee and Yalkowsky, 1988; Cline *et al.*, 1991).

Solubility is generally controlled by the polarity effect, which decreases in size for molecules with the same organic function. Non-polar or weakly polar substances dissolve in similar solvents. Thus, highly polar compounds dissolve in polar solvents such as water. The polarity or dipolar moment is proportional to the dielectric constant, and therefore high dielectric constant compounds (values of 80 for water and 34 for methanol) dissolve ions through hydration of the disassociated types (Fernandez and Quigley, 1985).

On the surface of clay-minerals, the absorbed water forms a double layer, which reduces the strength of interaction between the negatively charged clay particles and the cations in the colloidal solution. The hydrophobic hydrocarbons in the gasoline have low dielectric constant values, thus provoking the collapse of the double layer. This collapse is due to the contraction of the double layer through the attraction of the contra-ions which are closer to the superficial charge of the clay-minerals, favoring flocculation, and consequently the increase in permeability due to the increase in pore space (Mesri and Olson, 1971; Fernandez and Quigley, 1985 and 1988).

The co-solvêncy is responsible for the partition of BTXs to the aqueous phase, promoting the reduction of density of colloidal solution of soil, providing increased viscosity and a reduction of the surface tension (Mcdowell and Powers, 2003). This reduction in surface tension and generated by the collapse of the electrical double layer that there was between the soil and water (Farias, 2003).

Aspects of the Transport of Pollutants in Soils

The transport of pollutants in the soil can occur through the porous medium and saturated or unsaturated fractured media. This transportation occurs through physical or chemical processes, or through both processes. The chemical process becomes evident when the velocity of the fluid is not sufficiently high (i.e., less than 10^{-6} cm/s), generating a gradient due to the flow of the solute (contaminating agent) from the more concentrated medium to the less concentrated one. This process is called molecular diffusion (Rowe, 1988; Pastore and Mioto, 2000). This type of flow has been widely studied with metals and organic compounds in solid waste landfill leachate contaminants, for application in compacted soil layer, also called liners (Shackelford and Daniel, 1991; Rowe, 1988; Barone *et al.*, 1988).

Fernandez and Quigley (1985) developed an experimental research program to evaluate the hydraulic behavior of clayey-like soil (Sarnia, Ontario), permeated with liquid substances such as benzene, xylene, cyclohexane, aniline, propanol, acetone, alcohol and water. The results have shown that Hydraulic conductivity increased from 5 x 10^{-9} to 1 x 10^{-4} cm.s^{-1} along with a decrease in the dielectric constant from 80 (water) to 2 (benzene).

When there is a hydraulic gradient, the velocity of the solvent is relatively high and the transportation of the solute is practically managed by the velocity of the solvent, a mechanism which is known as an advection process. In this process, the velocity of the fluid is governed by Darcy's Law, which considers not only the characteristics of the soil, but also those of the fluid (Fernandez and Quigley, 1988).

In order to have good performance, the compacted clay liners must have a hydraulic conductivity less than 10^{-8} cm/s. However Daniel and Koerner (1995) defined that the hydraulic conductivity of clay liners must be less than or equal to 10^{-7} cm/s. This low flow is normally associated with the presence of clay-minerals, and at least 15 to 20% of particles with sizes under 2 mm, as well as a minimum plasticity greater than 7%, activity greater than 0.3, and cation exchange capacity (CEC) greater than 100 mmol$_c$/dm^3 of soil (Rowe *et al.*, 1995).

The natural organic material of the soils have proven to be efficient in the retarding process through the sorption of hydrophobic hydrocarbons, which are also found in gasoline (Chiou *et al.*, 1983; Karickhoff *et al.*, 1979; Schwarzenbach *et al.*, 1993).

Importance of Research

The aim of this study is to evaluate the behavior of a tropical soil, and their performance as liner against the flow of hydrocarbons from gasoline, by interpreting transportation according to physical and chemical parameters, as well as micromorphological aspects. For this characterized the mineralogy of the soil and the influence of his organic matter (OM), considering the adsorption processes of hydrocarbons from gasoline and hydraulic behavior in the laboratory by variation of the hydraulic gradient in front of the gasoline flow through compacted soil. This study also aims to contribute to the understanding of the dynamics of the flow through the soil of specific groups of compounds: aromatic, olephine, saturated hydrocarbons and the ethanol found in Brazilian type-C gasoline (a complex mixture of organic compounds).

LOCATION AND SOIL CLASSIFICATION

The soil sample was collected indisturbed block in depth of 4 m in the experimental field of foundations and test field of the Civil Engineering Department of the University of Brasília, located on the University campus in the city of Brasília, Brazil with coordinates 15° 56 ' 45 "S, 47° 52 ' 20" W (Fig. 1).

The sample of lateritic soil typical of the Brazilian *cerrado* region was studied. According to the Brazilian Soil Classification System (Embrapa, 1999), the soil was classified as Red Latossoil, considered as Ustic Rhodic Oxisol according to the U.S. Soil Taxonomy and Geric Ferralsol Ferric (FAO, IUSS Working Group WRB, 2007). It possesses a silt-clay-like texture, a large quantity of granular aggregates, and small pores. Visually, it is homogeneous and isotropic, without the presence of discontinuities.

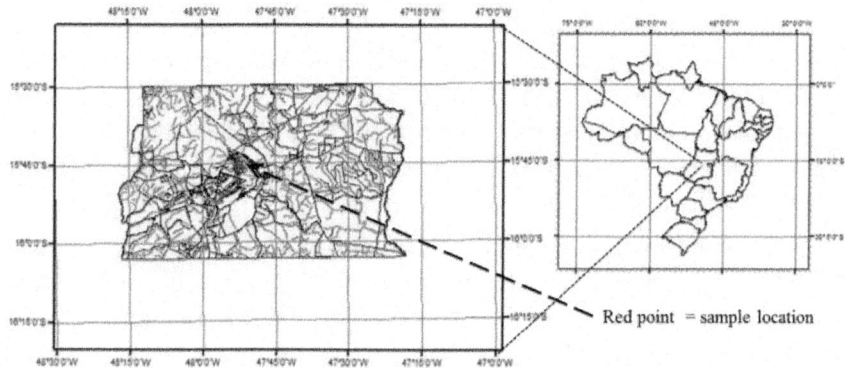

Figure 1: Map of location of the soil collection.

METHODOLOGY

The characterization of the soil involved the use of physical, chemical and mineralogical analysis.

Physical Tests

Geotechnical tests of physical properties of soils were performed following the Brazilian Association of Technical Standards (ABNT) procedures: test of limits of consistency called Atterberg limits following the ABNT NBR 7180/84 plastic limit; 6459/84 liquid limit following the Casagrande method. Before the grain-size determination, the real density was determined according the ABNT NBR 6508/84 method. The grain-size distribution curve was determined using a grain-size digital meter Malvern Mastersizer with lens de 300Rf for grain size range of 0.05 μm to 900μm at 25 ° C. For this analysis, the sample was previously passed through a No. 40 sieve. The analyses of the samples were done either with or without ultrasound dispersion. Ultrasonic condition was 5 minutes of dispersion in distilled water with ultrasonic level set at 5. The grain size fractions were classified following the Brazilian standard NBR 6502/93.

The degree of flocculation and dispersion of soil particles was determined comparing the results of grain size determinations before and after ultrasonic dispersion.

Hydraulic Conductivity

The test of hydraulic conductivity in compacted soil in standard Proctor energy were performed in a conventional manner with water using the variable charge and special form for gasoline (Fig. 2 and Fig. 3).

Figure 2: Hydraulic Conductivity cell.

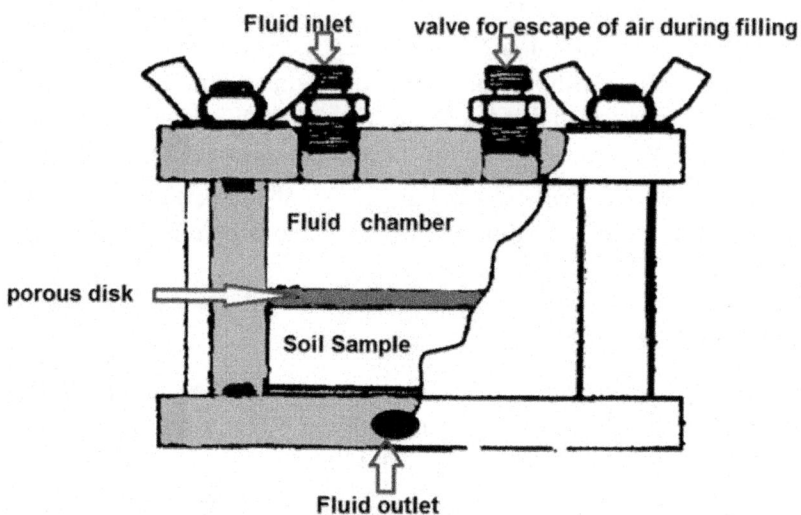

Figure 3: Schematic of permeameter Cell.

The gasoline hydrocarbons, for possessing volatile and low-density compounds require a special sealed cell to avoid losses due to evaporation and leakage and to support the applied tensions. The material selected for the construction of the special cells was stainless steel, to avoid reaction and adsorption problems in the walls, which is the case of plastics and acrylics (Doanhue *et al.*, 1999).

The system used to perform the gasoline's hydraulic conductivity was similar to that applied by Fernandes (1989). The special cell may be disassembled, and is made up of three parts. The first part is a cylinder, where the test material and reservoir are found. This part is 5 mm thick, 110 mm long and has an internal diameter of 77.2 mm. The other two parts are the upper and lower lids. Both have cavities filled with rubber rings which are able to prevent the reaction of the hydrocarbons in the gasoline and act as a seal when the cylinder is assembled. The upper lid has two openings, one for the entry of fluid and the other for the application of vertical tension with compressed air. The lower lid is made up of an outgoing flow register which is connected to a collecting container. The two lids are $120 \times 120 \text{ mm}^2$ square, 10 mm thick. The connections were made out of aluminum, due to its low cost and flexibility; the connecting joints were sealed with 3M automotive glue and winding sealing thread, in order to prevent leaks and to make the system more secure.

The conductivity test was performed with test material 5 cm long, compacted at normal Proctor energy at optimal water content condition, in the cylinder of the hydraulic conductivity cell. Then, a thin disk of porous ceramic was placed top of the sample. The small space between the disk and the cylinder wall was filled with 3M glue to prevent preferential flows along the wall, and to ensure that the gasoline only passed through the porous ceramic disk. The cell was then assembled, and the upper and lower lids were connected to the cylinder. The cylinder is 11 cm high, of which the remaining 6 cm were filled with type C gasoline. After the cell was totally sealed and connected to the compressed air system, with pressure controlled by a manometer, it was connected with plastic tubes able to support high pressure. The conductivity tests were performed for various applied vertical pressures. For each pressure applied, the hydraulic conductivity was measured. The pressures were varied to see how the soil sample behaved with an increase in hydraulic gradient upon the flow of gasoline. The hydraulic conductivity was measured in the laboratory at static tensions σ_v of 50, 100, 150, 200, and 300 kPa, with respective hydraulic gradients of 75, 150, 225, 300, and 450.

The residual water of the soil pores mixed with gasoline collected in the test was previously run through a separating funnel to remove the aqueous phase to later take a reading of hydrocarbons of the gasoline through infrared technique.

The test material of the lateritic soil sample, before and after the hydraulic conductivity test conducted with water, and the other with the flow of gasoline, were dried at room temperature. Micromorphological analyses were performed on Thin Lamina (TL) in vertical sections, prepared by impregnating the sample with plastic resin (Cardoso, 1995; Martins, 2002). The instrumental technique used for the microscopic views of the TL was Optical Microscopy.

Mineralogical Characterization

The identification and quantification of minerals in the sample were carried out by the method developed by Martins (2000). This method involves the use of X-ray diffraction (XRD) technique for identifying the minerals, chemical analysis for the determination of major elements (Al, Fe, Si, Ca, Mg and Ti), thermogravimetric analysis (TGA), and the use of Munsell color code (Munsell color company Inc., 1975). The determination of major chemical elements was performed by ICP-AES after the fusion of samples with alkaline NaOH as fondant at a temperature of 450 ° C for 40 minutes using the nickel crucible. Determinations of elements by ICP / AES (atomic emission spectrometry of Plasma Induced Coupling) were performed with Thermo Jarrell ASH equipment, model Iris / AP.

The thermogravimetric analysis were applied to quantify the kaolinite and gibbsite. For this used the TGA Shimadzu equipment with temperature ramp of 20 °C to 1500 °C, with speeds ranging from 0.2 to 60 ° C / min, using the software TAS60WS for the treatment of data. The Munsell code was used for determining the ratio of hematite and goethite in the soil samples. The CEC of soil was determined using the principle of the simple as the sum total of the exchangeable cations that a soil can adsorb. The determination of the organic matter content was done prior to extraction using wet oxidation method with potassium dichromate in sulfuric medium. The excess of dichromate after oxidation was titrated with standard solution of ferrous ammonium sulfate (Mohr salt).

Chemical Characterization

The pH was measured in the soils samples in distilled water medium using a combined glass electrode Ag/AgCl (potentiometric method).

In order to study the influence of OM and mineralogy in the gasoline sorption process, an experiment was performed with samples treated with H_2O_2 and another without treatment.

The extraction of the OM used 15 g of soil in a porcelain capsule, with 10 mL of H_2O_2 volume 30% and with agitation in a 50 mL Becker cup. After agitation, there was an effervescent reaction, when the capsule was covered with clock glass for one night. The process was repeated until the complete disappearance of the reaction. It was then washed 3 to 5 times in distilled water, using a Büchner funnel with filtering under reduced pressure. Then, for the gasoline sorption test, the sample was allowed to dry at room temperature.

The sorption test procedure used 2 g of soil with 25 mL of gasoline placed in an amber glass jar under agitation for 24 hours at a temperature of 22°C. After this, the samples were centrifuged as in the processes described above, with the removal of a 15 mL portion for analysis.

The hydrocarbons content of the gasoline samples was determined at the National Petroleum Agency (ANP) laboratory, in Brasilia, with a (FTIR = Fourier Transform Infrared), manufactured by Grabner Instruments, model IROX 2000. This instrument qualified and quantified the compounds, generating the mass and volume percentages of the ethanol, aromatic, olephine and saturated compounds.

RESULTS

Tab. 1 and 2 present data of the physical, chemical and mineralogical Brazilian soil and constituents of the gasoline type C studied.

Table 1: Characteristics of the soil (Farias, 2003).

Test	Lateritic
Atterberg Limits	
Liquid limit-W_L (%)	41
Plastic limit-W_p (%)	29
Plastic Index-I_p (%)	12
Activity	0,18
Grain size distribution*	
Clay (%)	65
Silt (%)	34
Sand(%)	1
Degree of flocculation (%)	92
Degree of dispersion (%)	8
Chemical Parameters	
pH	5,70
Organic Matter content (%)	0,41
CEC ($mmol_c/dm^3$)	6,4
Mineralogy	
Quartz (%)	30,2
Anatase (%)	1,57
Kaolinite(%)	24,6
Gibbsite (%)	25,5
Goethite (%)	4,6
Hematite (%)	7,5
Illite (%)	2,2
Vermiculite (%)	3,7
Hydraulic Conductivity in water (cm/s)	3,7.E-07

[i] - *Grain size data obtained by ultra-sound waves using a laser beam grain size analyser.

Tab. 2 presents the composition of the Brazilian type-C gasoline, according to Farias (2003).

Table 2: Brazilian Type C gasoline data

Compounds	Mass (%)
Aromatics	20,8
Olefins	22,4
Saturated	31,4
Ethanol	25,4

Fig. 4 presents the increase in hydraulic conductivity with an increase in the hydraulic gradient. At a gradient of approximately 210, conductivity becomes practically constant. Fig. 5 presents the intrinsic permeability, which considers the characteristics of the soil, but does not consider the chemical and physical properties of the fluid. Intrinsic permeability reaches values close to $10^{-13}m^2$. However, as the hydraulic gradient increases, stability reaches approximately $10^{-11}m^2$.

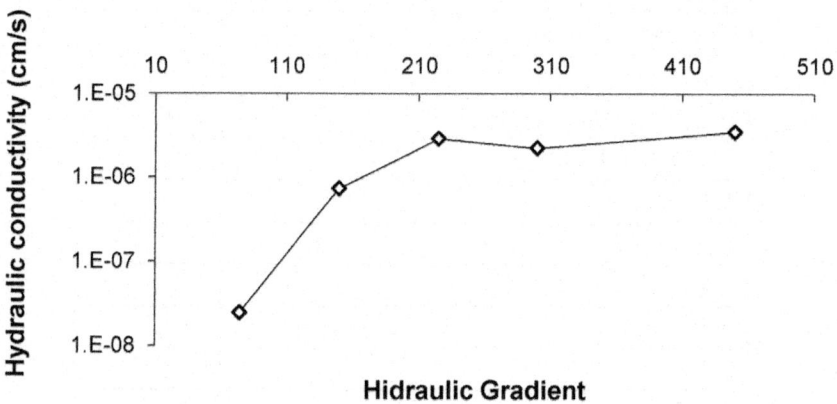

Figure 4: Behavior of hydraulic conductivity and hydraulic gradient of laterite soil on the gasoline flow.

Fig. 6 depicts the behavior of the hydraulic conductivity relative to the volume of pores while undergoing saturation in the test material with gasoline at a tension of σ_v of 50 kPa. The saturation process takes place with the expulsion of the interstitial water accumulated in the pores due to optimal compacting moisture content ($w_{opt} = 26\%$) is the test material at normal Proctor energy. It may be observed that as the volume of pores in the gasoline flow increases, conductivity decreases from 4 to 2×10^{-8} cm.s^{-1}. This suggests that the behavior of the reduction may be represented by a second-order equation.

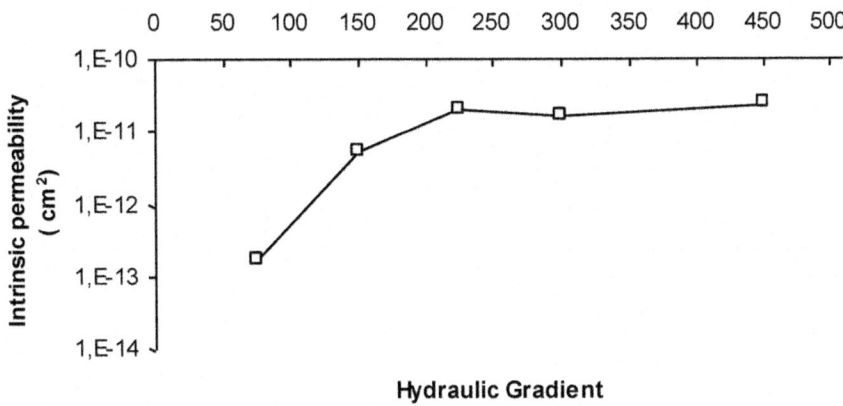

Figure 5: Behavior of the intrinsic permeability and hydraulic gradient of laterite soil on the gasoline flow.

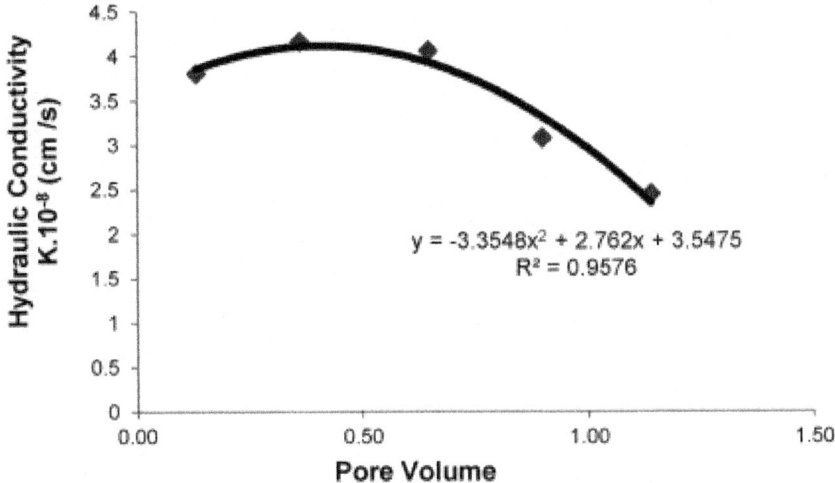

Figure 6: Behavior of the lateritic soil saturated with gasoline at 50 kPa.

Fig. 7 presents the saturation process at a σ_v tension of 50 kPa, based on the ratio between the concentration (C) of the gasoline hydrocarbons passing through the soil sample, and the initial concentration (C_o) added to the reservoir, in relative to the volume of pores. The hydrocarbons concentration data are from the Light Non-aqueous Liquid Phase (LNALP), after the flow through the soil sample in the hydraulic conductivity test.

Table 3: Result of the physical parameters of the test material

Sample	w (%)	γ (kN.m^{-3})	γ_{dmax} (kN.m^{-3})	γ_s (kN.m^{-3})	e	n	S r (%)	Vv c^{m}3
lateritic*	1,7	17,7	17,4	27,5	0,58	0,4	8,1	134,3
lateritic**	1,7	15,8	15,6	27,5	0,77	0,4	6,2	178,5
lateritic***	1,8	14,7	14,5	27,5	0,90	0,5	5,3	210,0

[i] - *Dry soil sample before the hydraulic conductivity test

[ii] - **Dry soil sample after the hydraulic conductivity test with the water flow

[iii] - *** Dry Soil sample after the hydraulic conductivity test with the gasoline flow

The results in Tab. 3 present the physical parameters of the compacted test materials dried at room temperature before and after the hydraulic conductivity test. Highlights the volume of voids (**Vv**), which changes substantially when there is a flow of gasoline. The degree of saturation also decreases after the flow of gasoline.

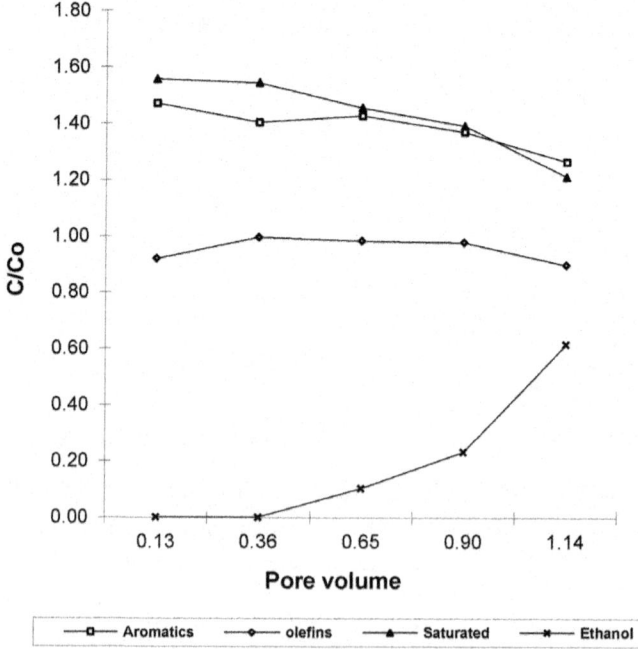

Figure 7: Light non-aqueous liquid phase ratio of the gasoline relative to the volume of pores of the lateritic soil in a saturation process at 50 kPa.

The micromorphology of the three compacted soil samples was important in order to visualize the behavior of the test material before the hydraulic flow (Fig. 8), after the hydraulic flow with water, and after the flow with gasoline. It must be noted that the grains of quartz make up approximately 40% of the total solid material; variable in size, 0.12 mm on average; and overall, are sub-rounded to angular. They are highly fractured, without orientation and their contours present corrosion. In spite of the compacting, the structure of this soil is not totally dispersed, for microaggregations of oxyhydroxides of Fe and Al remain, forming micropores. The compacted soil sample submitted to percolation in water showed a single micro-structural difference relative to the one performed on the LT of the compacted soil sample. Actually, there was an increase in small canal-type voids, generated by the flow of water (Fig. 9). The micromorphology regarding the LT of the compacted soil submitted to the flow of gasoline also showed only a quantitative increase in canal-type voids (Fig. 10). However, this variation was greater than that registered in the previous sample with the water flow.

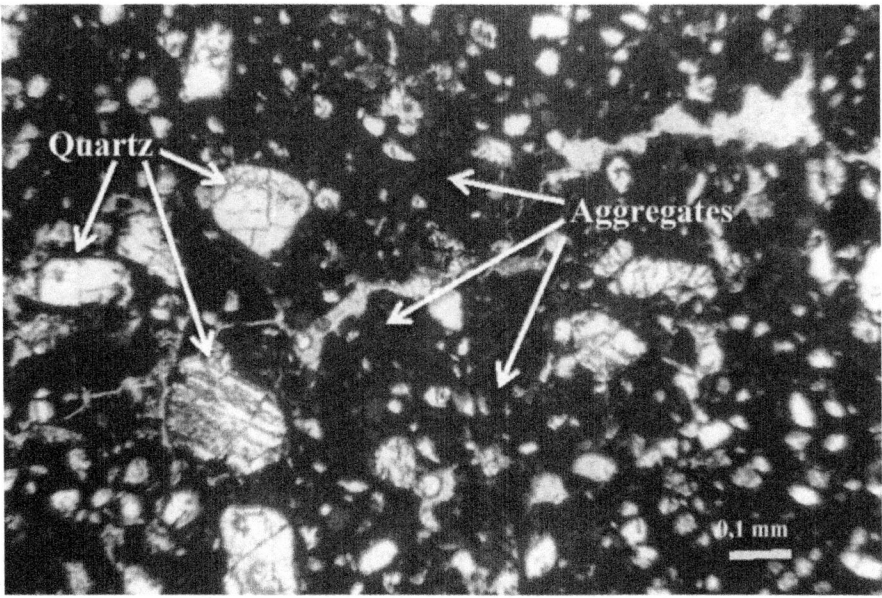

Figure 8: Photomicrography of the porfirosquelic APE, aggregates, and quartz grains of the compacted lateritic soil. Parallel nichols (N//).

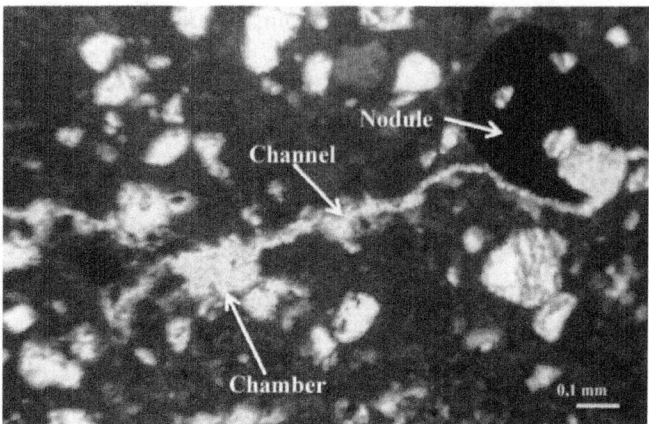

Figure 9: Photomicrography showing the nodules and canal- and chamber-type voids of the compacted lateritic soil submitted to percolation with water. Parallel nichols (N//).

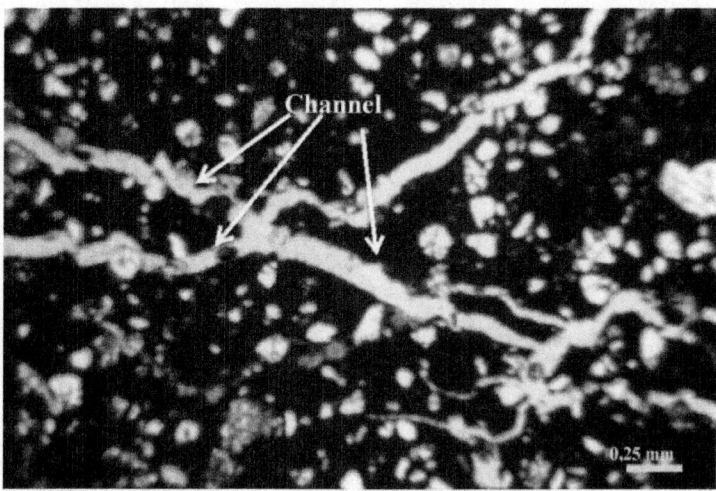

Figure 10: Photomicrography showing the canal-type voids of the compacted lateritic soil submitted to percolation with gasoline. Parallel nichols (N//).

Fig. 11 presents the results of the adsorption of the ethanol and aromatic substances in the samples with and without the extraction of organic matter with the use of hydrogen peroxide. Note that the samples treated with extractor presented low adsorption. Aromatic compounds showed no adsorption after extraction of organic matter contained in the soil.

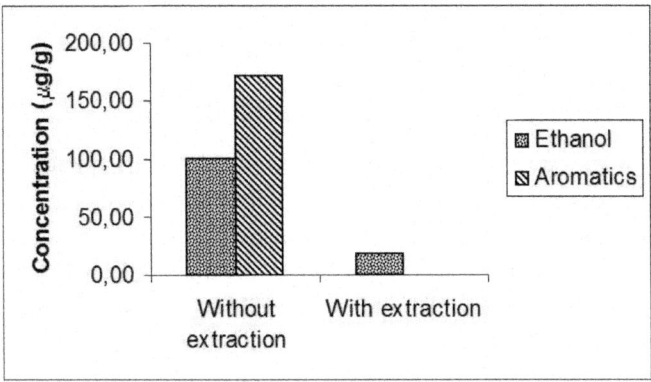

Figure 11: Results of the adsorption of the gasoline hydrocarbons in the soils with and without the extraction of the soil organic matter.

Gasoline ethanol can be adsorbed on the sites of hydroxyls of the octahedron of Al, exposed by fractures, Scrubs or crystalline lattice imperfections, or by interactions with the Fe oxides and hydroxides and Al amorphous. This occurs from adsorption of hydrogen bonds, which can also occur with water strongly adsorbed on the surface of the clay minerals (Fig. 12).

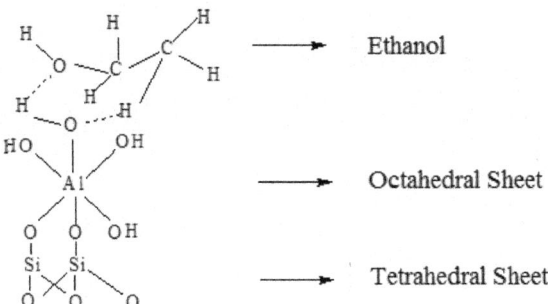

Figure 12: Coordination of interaction of hydrogen and hydroxyl ethanol exposed in the clay mineral (1:1).

DISCUSSION OF THE RESULTS

The discussion of the results is focused on three main aspects. The first considers mineralogical, chemical and physical characteristics of the material with potential for liners. The second aspect is assessed the performance of Laterite soil on gasoline hydrocarbon flow subjected to high hydraulic gradients, causing an acceleration of the process of formation of flow channels for compressed soil in power of Proctor. The last aspect to be evaluated is the power of gasoline hydrocarbon adsorption by soil with OM and no OM.

The Delimiters Criteria Material with Potential for Liners

Evaluating the criteria prescribed by Rowe et al., (1995) the soil presents considerable levels of Fe oxides and hydroxides and Al (hematite, goethite and gibbsite) and kaolinite with only 30.2% of quartz. As the mineral vermiculite is low soil activity levels was 0.18, less than the 0.3 suggested by the literature. However this value of activity indicates that the material is not expandable, being a good quality for liners. The cationic exchange capacity (CEC) also presented low value (6.4 $mmol_c/dm^3$) comparing with value defined in literature. Tropical soils lateritic in general are highly weathered with low or no mineral content of 2:1, which are typical of temperate climate. Therefore, the activity and CEC are low. The granulometry performed 65% clay fraction indicated more than 20% of particles less than 2 mm confirming the material rich in clay fraction indicating low permeability material when compressed. Thus, the hydraulic conductivity parameter value introduced into water in the order 10^{-7}cm/sec subjected to a pressure of 20 kPa (Tab. 1). These results of the hydraulic conductivity characterize the material with great potential for liner according to predefined values in the literature.

Lateritic Soil Performance as Liner

The hydraulic conductivity of gasoline type C brazilian obtained values between 10^{-8} to 10^{-7} cm/s to a gradient of 75 with a pressure of 50 kPa, which corresponds to 5 m of the water column (Fig. 4). Such a result of hydraulic conductivity defines the material as great for barrier on gasoline hydrocarbon flow according to predefined values in literature (Rowe et al., 1995; Daniel and Koerner, 1995). With the increase of the hydraulic gradient there was an increase in hydraulic conductivity until it reaches a level of stabilization in gradient greater than 210 (Fig. 4). Although it does not occur to the destruction of the liner to the 210 to 450 gradients suggests avoid gradients greater than 100 in the projects of protection of underground fuel tanks, ensuring in this way a hydraulic conductivity around 10^{-7} cm/s for liners according to the literature. The intrinsic permeability or specific considers simply the porous medium, not considering the characteristics of fluid. The values found for intrinsic permeability compacted laterite soil is similar to that found in the literature to clay (Freeze and Cherry, 1979).

The compacted soil voids indexes suffered increased 0.58 before tests to 0.77 with water flow and 0.90 with gasoline flow in hydraulic gradient of 75. The empty volume also increased from 134.3 to 210.0 (Tab. 3). The soils studied presented a high degree of flocculation due to the aggregates of the oxyhydroxides of Fe and Al. Even when compacted, they contain micro-aggregates which are not destroyed. When a flow is established through the soil, the micro-aggregates may interconnect, forming flow channels. The physical behavior provoked by the flow may be visualized in the micromorphology of the samples in Fig. 8, 9 and 10. However even with these micros channels formed in the compacted soil hydraulic conductivity limit of 10^{-7} cm/s (Daniel and Koerner, 1995) is not affected considering a gradient of 100.

Adsorption Performance for Hydrocarbons of Gasoline

The performance of laterite soil to a gasoline hydrocarbon flow subjected to a pressure of 50 kPa with 75 gradient was evaluated for pore volume and the ratio C/Co in the process of saturation of the compacted clay liner for gasoline, in other words, there was the expulsion of the water contained in the soil by the process of compression to achieve the optimum water content of compaction. The reason indicates that values above 1 there is an LNAPL phase concentration of groups of substances evaluated. The groups were evaluated for aromatic, olefins, saturated and ethanol.

In Fig. 7, the aromatic compounds appear as constants in the saturation process. Since they are hydrophobic, their polarity is low, and are more easily transported in the soil. The olefines and saturates have a greater C/Co ratio in the LNAPL due to their low solubility in water, being lower than the aromatic compounds, which are more affected by ethanol through co-solvency. In the 0.13 a 0.36 pore volume range, the ethanol is partitioned to the aqueous phase and, as the saturation of the pores with gasoline increases, the C/Co ratio for ethanol in the LNAPL also increases. The partitioning of the ethanol for the aqueous phase is natural and is due its polarity, which makes it mixable in water. Thus, the ethanol, along with the other hydrophobic compounds in the gasoline, favors the collapse of the double layer, as well as the increase in micropores (Rowe et al., 1995)

The results in Fig. 11 show that the soil organic matter, although in low quantities, has an influence of almost 0.41 %, in the sorption process. The soil studied was collected at a depth of 4 meters, thus contained evolved organic matter, possibly fulvic acid. The removal of the organic material with hydrogen peroxide showed low ethanol sorption. The aromatic compounds, which are hydrophobic, were not absorbed.

Evaluating the transportation of gasoline compounds by soil with application of 50 kPa of pressure, indicates low retention and greater mobility, because mostly they are hydrophobic compounds that do not bind the soil particles. Another aspect of this experiment is that has not been evaluated by diffusion flux, which occurs at speeds equal to or less than 10^{-10} cm/s. The test of sorption to organic matter proved to be important in the retention process of ethanol. In view of the low adsorption of gasoline compounds by soil suggests considering projects of liners gradients below 75 and pressures less than 50 kPa ensuring a hydraulic conductivity greater than 10^{-8} cm/s and use clayey material rich in organic matter to promote greater retention of ethanol and avoid or reduce the effect of co-solvency.

CONCLUSION

Since the lateritic soil studied possesses a high aggregation capacity, even when compacted at normal Proctor energy, micropores remain which, in high hydraulic gradient situations, are interconnected, forming flow channels. However, even under higher hydraulic gradients in the gasoline percolation tests, this soil presents good material of liners. This is due to the stabilizing of the flow channels formed, favoring also the stabilizing hydraulic conductivity.

The measure that gasoline occupies the pores in the process of saturation the concentration of ethanol increases. This is due to the polarity of the ethanol. The aromatic compounds maintain a C/C_0 ration close to 1, as the volume of pores increases, indicating that these are tracers due to their low dielectric constant and polarity. Due to their low solubility in water, the olephines and saturates are more present in the LNAPL phase. These hydrocarbons may form emulsions, favoring transportation through the soil in the aqueous phase.

Regarding the retarding potential of the lateritic soil, evaluated by the sorption parameter, it is not directly correlated with the mineralogy, because the aromatic compounds are not absorbed when the organic material is extracted. Actually, this sorption may be correlated with a certain type of humic substance, which may be interacting with the poly-amorphs of the oxyhydroxides of Fe and Al in the soil, favoring interaction with the aromatic compounds.

Finally, a low hydraulic gradient context (< 75), hydraulic conductivity < 10^{-8} and organic matter, in lateritic soil can improve the performance of liner.

ACKNOWLEDGEMENTS

The authors are gratefully to the Conselho Nacional de Desenvolvimento Científico e Tecnológico – CNPq, CAPES and ANP for the fellowships and financial support granted to the accomplishment of this research.

REFERENCES

1. ABNT (1984) Solo – Determinação do limite de liquidez –NBR 6459/84. Associação Brasileira de Normas Técnicas, Rio de Janeiro, RJ, 6p.

2. ABNT (1984) Solo – Determinação do limite de plasticidade – NBR7180/84. Associação Brasileira de Normas Técnicas, Rio de Janeiro, RJ, 3p.

3. ABNT (1984) Solo – Grãos de solo que passam na peneira de 4,8 mm - Determinação da massa específica (método de ensaio) – NBR 6508/84. Associação Brasileira de Normas Técnicas, Rio de Janeiro, RJ, 8p.

4. ABNT (1993) Rochas e solos – Terminologia – NBR 6502/93. Associação Brasileira de Normas Técnicas, Rio de Janeiro, RJ, 19p.

5. Banerjee, S. and Yalkowsky, S.H. (1988). Cosolvent-Induced Solubilization of Hidrophobic Compounds into Water. Analytical Chemistry, v.60, p. 2153-2155.

6. Barone, F.; Yanful, E.K.; Quigley, R.M. and Rowe, R.K. (1988). Effect of multiple contaminant migration of diffusion and adsorption of some domestic waste contaminants in a natural clayey soil. Geotechnical Research Report GEOT -5-88, Geotechnical Research Centre. The University of Westem Ontario, London, Ont.

7. Cairney, S., Maruff, P., Burns C., Currie B (2002) The neurobehavioural consequences of petrol (gasoline) sniffing. Neuroscience and Biobehavioral Reviews, v.26:1, p. 81-89.

8. Cardoso, F. B.F (1995) Análise Química, Mineralógica e Micromorfológica de Solos Tropicais Colapsíveis e o Estudo da Dinâmica do Colapso. Dissertação de Mestrado, Departamento de Engenharia Civil, Universidade de Brasília, Brasília, DF, 142p.

9. Chiou, C.T., Porter, P.E. and Schmedding, D.W. (1983). Partition equilibrium of nonionic organic compounds between soil organic matter and water. Environmental Science and Technology, 17(4):227-231.

10. Cline, P.V., Delfino, J.J. and Rao, P.S.C. (1991). Partitioning of aromatic constituents into water from gasoline and other complex solvent mixtures. Environmental Science and Tecnology, 25(5):914-920.

11. Corseuil, H. X. ; Fernandes, M.(1999). Efeito do Etanol no Aumento da Solubilização de Compostos Aromáticos Presentes na Gasolina Brasileira. Revista Engenharia Sanitária e Ambiental, v. 4, n. 1 e 2, p. 71-75.

12. Daniel, D.E and Koerner, R.M. (1995) Waste containment facilities – Guidance for construction, Quality Assurance and Quality Control of

Liner and Cover Systems, American Society of Civil Engineers, ASCE Press, New York, 354 pp.

13. Donahue, R.B., Borbour, S.L. and Headley, J.V. (1999). Diffusion and adsorption of benzene in Regina clay. Canadian Geotechnical Journal, 36(3):430-442.

14. FAO, IUSS Working Group WRB (2007). World Reference Base for Soil Resources 2006, first update 2007. World Soil Resources Reports No. 103, Rome.

15. Farias, W.M. (2003). Condutividade Hidráulica de Solos Tropicais Compactados a Hidrocarbonetos da Gasolina. M.Sc. thesis, Department of Civil and Environmental Engineering, University of Brasilia, Brasilia DF, 152 pp.

16. Fernandez, F. and Quigley, R.M. (1985). Hydraulic conductivity of natural clays permeated with simple liquid hydrocarbons. Canadian Geotechnical Journal, 22:205-214.

17. Fernandez, F. and Quigley, R.M. (1988). Viscosity and dieletric constant controls on the hydraulic conductivity of clayey soils permeated with water – soluble organics. Canadian Geotechnical Journal, 25:582-589.

18. Freeze, R. A. and Cherry, J. A (1979). Groundwater. Prentice-Hall, New Jersey, USA, 604p.

19. Karickhoff, S.W., Brown, D.S. and Scott, T.A. (1979). Sorption of hydrophobic pollutants on natural sediments. Water Research, 13:241-248.

20. Martins, E.S., (2000). Petrografia, Mineralogia e Geomorfologia de Rególitos latériticos do Distrito Federal. Tese de Doutorado, Instituto de Geociências, Universidade deBrasília, Brasília, DF, 196 p.

21. McDowell, C.J. and Powers, S.E. (2003). Mechanisms Affecting the infiltration and distribution of ethanol-blended gasoline in the vadose zone. Environmental Science and Technology, 37:1803-1810.

22. Mesry, G. and Olson, R.E. (1971). Mechanisms controlling the permeability of clays. Clay and Clay Minerals, 19(3):151-158.

23. Mussell Soil Color Company (1975) Mussell soil color charts. Baltimore, 1V, 117p.

24. Pastore, E.L. and Mioto, J.A. (2000). Impactos ambientais em mineração com ênfase à drenagem mineira ácida e transporte de contaminantes. Solos e Rochas. São Paulo, SP, 23(1):33-53.

25. Rowe, R.K. (1988). Contaminant migration though groundwater: The role of modeling in the design of barriers. Canadian Geotechnical Journal, 25(4):778-798.

26. Rowe, R.K., Quigley, R.M. and Booker, J.R. (1995). Clayey Barrier Systems for Waste Disposal Facilities. E and FN Spon, London, England, 390p.

27. Scharzenbach, R.P., Gschwnd, P.M. and Imboden, D.M. (1993). Environmental Organic Chemistry. Wiley Interscience, New York, NY, 1313p.

28. Shackelford, C.D. and Daniel, D.E. (1991). Diffusion in saturated soil. II: Results for compacted clay. Journal of Geotechnical Engineering, 117(3): 485-506.

29. Ulrich, G. (1999). The Fate and Transport of Ethanol-Blended Gasoline in the Environment. Governors' Ethanol Coalition, Lincoln, Nebraska, 88p.

Chapter 3

SOIL MOISTURE ESTIMATION IN SOUTH-EASTERN NEW MEXICO USING HIGH RESOLUTION SYNTHETIC APERTURE RADAR (SAR) DATA

A.K.M. Azad Hossain [1] and Greg Easson [2]

[1]Department of Geology and Geological Engineering, The University of Mississippi, 120A Carrier Hall, University, MS 38677, USA

[2]Mississippi Mineral Resources Institute, The University of Mississippi, University, MS 38677, USA

ABSTRACT

Soil moisture monitoring and characterization of the spatial and temporal variability of this hydrologic parameter at scales from small catchments to large river basins continues to receive much attention, reflecting its critical role in subsurface-land surface-atmospheric interactions and its importance to drought analysis, irrigation planning, crop yield forecasting, flood protection, and forest fire prevention. Synthetic Aperture Radar (SAR) data acquired at different spatial resolutions have been successfully used to estimate soil moisture in different semi-arid areas of the world for many years. This research investigated the potential of linear multiple regressions and Artificial Neural Networks (ANN) based models that incorporate different geophysical variables with Radarsat 1 SAR fine imagery and concurrently measured soil moisture measurements to estimate surface soil moisture in Nash Draw, NM. An artificial neural network based model with vegetation density, soil type, and elevation data as input in addition to radar backscatter values was found suitable to estimate surface soil moisture in this area with reasonable accuracy. This model was applied to a time series of SAR data acquired in 2006 to produce soil moisture data covering a normal wet season in the study site.

INTRODUCTION

Soil moisture, is an important hydrologic variable that controls the interactions (and feedbacks) between land surface and atmospheric processes [1]. It plays a very important role in the distribution of precipitation between runoff and infiltration. Soil moisture monitoring and characterization of the spatial and temporal variability of this hydrologic parameter at scales from small catchments to large river basins continues to receive much attention, reflecting its critical role in subsurface-land surface-atmospheric interactions and its importance to drought analysis, crop yield forecasting, irrigation planning, flood protection, and forest fire prevention [2,3,4].

In semi-arid environments ground water recharge is one of the most difficult parameters to quantify, where a number of recharge mechanisms, including soil moisture change, on variable temporal and spatial scales. Several studies showed that temporal analysis of soil moisture can be used to understand ground water recharge [5,6]. Remote sensing technology has been used successfully to estimate soil moisture [3,7,8,9,10] and map its spatio-temporal distribution in semi-arid environments and could potentially contribute to ground water recharge studies.

Synthetic Aperture Radar (SAR) data are particularly well suited for estimating soil moisture due to the relationship between the dielectric constant and soil moisture [11,12]. The microwave measurements are strongly dependent on the dielectric properties of the target. For a soil, dielectric properties are a function of the amount of water present. The real part of the complex dielectric constant of water in this spectral region (microwave) is approximately 80 comparing to the value for dry soil is about three. This large contrast provides a basis for estimating the moisture for dielectric values between these two extremes [13,14]. It is not always possible to take advantage of the dielectric constant and soil moisture relationship since microwave measurements are also influenced by surface roughness, which varies significantly from place to place due to diverse land use and land covers. That's why no global SAR based operational algorithm exists for estimating soil moisture [15].

Studies, particularly in the past two decades, have resulted in a multitude of methods, algorithms, and models relating satellite-based radar backscatter imagery to estimate surface soil moisture content [3,10,11,16,17,18,19,20]. The most commonly used algorithms are developed by a semi-empirical approach [21,22]. In a given bare soil condition, radar backscatter is linearly dependent on volumetric soil moisture content (θ_w) in the upper 2 to 5 cm of soil with a correlation (R^2)~ 0.8 to 0.9 [21,22]. However, the presence of vegetation cover complicates soil moisture estimation due to the interaction of the microwaves with the vegetation and soil [12]. The radar backscatter from a surface with vegetation consists of three components: (1) product of the backscatter contribution of bare soil surface ($\sigma_s°$) and the two way attenuation of the vegetation layer (τ^2), (2) the direct backscatter contribution of the vegetation layer ($\sigma°_{dv}$), and (3) multiple scattering involving the vegetation elements and the ground surface ($\sigma°_{int}$) [12].

$$\sigma° = \tau^2\sigma_s° + \sigma_{dv}° + \sigma_{int}°$$ (1)

In soil moisture estimation for semi-arid environments using different SAR data the influence of sparse vegetation were found negligible by several studies [11,16,17,18,19,22]. That means, for a given soil with uniform surface roughness (R), θ_w can be estimated using the following simple linear regression expression since in this case $\sigma° \approx \sigma_s°$.

$$\theta_w = a + b\sigma°$$ (2)

where a and b are regression coefficients, usually determined from field experiments encompassing the target-invariant, scene-invariant SAR wavelength, incidence angle, polarization, and calibration.

Using high resolution SAR data, however, it is not always possible to obtain strong linear relationship between measured soil moisture and radar backscatter even in semi-arid environments. Depending on the amount of vegetation present, its dielectric properties, height and geometry, the sensitivity of microwave backscatter to variations in volumetric soil moisture may be significantly reduced [23]. As shown by [23], in a semi-arid environment of south-eastern New Mexico vegetation coverage can significantly reduce the accuracy of soil moisture estimation using numerical models based on simple linear ($R^2 = 0.05$ to 0.24) and non-linear ($R^2 = 0.24$) relationships between radar backscatter from high resolution SAR imagery and near real time *in situ* soil moisture measurements. In addition to the vegetation density, the SAR backscattering can also be influenced by the variation in soil types and soil salinity since they may affect the surface dielectric properties. Topographic variations can also influence the distribution of soil moisture in the field.

This paper investigated the potential of multiple linear regressions and Artificial Neural Networks (ANN) based models to improve soil moisture estimation in south-eastern New Mexico using high resolution Radarsat 1 SAR imagery. The models used SAR backscatters and near real time soil moisture measurements along with vegetation density, soil type, elevation, and soil salinity measurements. A time series of SAR based soil moisture estimation data were generated for an entire wet season in the study site using the developed numerical models.

STUDY SITE

A small area in south-eastern New Mexico called Nash Draw was selected as the study site for this study (Figure 1). It is located about 30 km east of Carlsbad, NM. It covers an area of about 400 km²and the study site occupies 225 km² of it. The extent of the site is limited between 103.78°W–103.92°W longitude and 32.23°N–32.36°N latitude. It is a part of the north-eastern Chihuahua desert, which is characterized by semi-arid environments and sparsely vegetated rangeland. The topographic relief in the region is not significant. The maximum relief across the area is approximately 200 m.

Nash Draw is a karst valley that developed in response to subsurface dissolution of evaporites (including halite and sulfate rocks) and subsidence of the overlying strata [24]. It is a complex example of the localized effects of evaporite karst on surface topography, near-surface geology, and hydrology [25]. Different areas of Nash Draw display small karst features, including caves, sinkholes, dolines, and larger integrated forms such as valleys or elongated depressions [25].

The hydrologic system within Nash Draw is poorly understood. Much of Nash Draw exhibits no significant integrated surface drainage [25], while much of the area is covered by a thick blanket of dune sand. That is why it is very difficult to identify the potential locations for ground water recharge in this area. Therefore, to assist in understanding the existing hydrologic processes modifying Nash Draw it is very important to determine the spatio-temporal distribution of surface soil moisture.

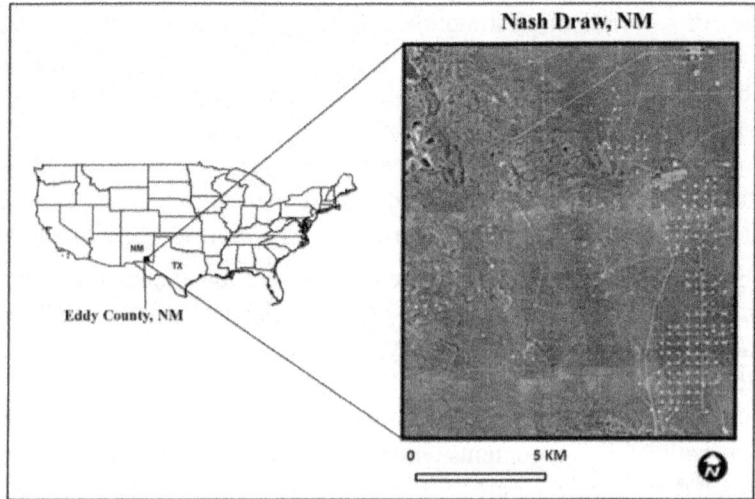

Figure 1: Location of study site.

Since the study site covers an area of 225 km² high resolution SAR data would be very suitable to obtain remote sensing derived soil moisture estimation for this area. This is the reason Nash Draw was found suitable as a test site for the conducted study.

Rainfall in Nash Draw is unreliable and erratic, with August as the wettest part of the year and the rainy season ending in October. Therefore, it was considered that imagery acquired from August to November should record the maximum variation of soil moisture in the study site.

DATA ACQUISITION AND PROCESSING

Synthetic Aperture Radar (SAR)

Due to its high spatial resolution, Radarsat 1 SAR Fine imagery was selected for estimating soil moisture in the study site. Five scenes of Radarsat 1 SAR Fine imagery, with 10 m spatial resolution and 50 km swath, covering the Nash Draw area were acquired from the Alaska Satellite Facility (ASF). The scenes were acquired in descending mode at a 37° incidence angle on 2 and 26 August, 19 September, 13 October, and 6 November of 2006. Figure 2 shows the Time series of the acquired SAR imagery.

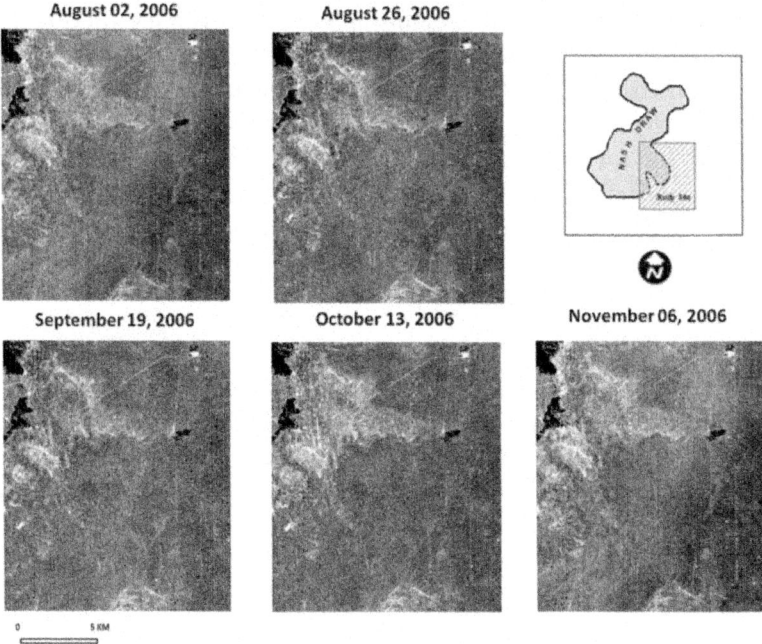

Figure 2: Time series of acquired SAR imagery.

SAR Pre-Processing

All acquired SAR imagery were received as Level 0 products and then converted to level 1 products. These Level 1 SAR data were terrain corrected, calibrated, and filtered in preparation for soil moisture estimation.

Data Calibration

Sigma-naught ($\sigma°$) describes the average reflectivity or scattering co-efficient of a radar scene. Beta-naught ($\beta°$) interprets the brightness estimates of mean reflectivity and separates the radiometric response and reflectivity dependent on the terrain properties, such as local slope. According to [26]:

$$\sigma^o = \beta^o + 10\log_{10}(\sin l) \tag{3}$$

$$\beta^o = 10\log_{10}(DN^2 + A_3/A_2) \tag{4}$$

where, DN—Digital number, A_3—fixed offset from the radiometric data record, A_2—Look-Up Table (LUT) value, and l—local incidence angle.

Calibrating the radar scene in sigma naught ($\sigma°$) requires normalization using the knowledge of local incidence angle as we can see in Equation (3). However, this normalization is not required for beta naught ($\beta°$) and for an area with low relief (that is, no significant influence with the variation of local incidence angle). Since the topographic relief is not significant in the study site the acquired SAR data were calibrated as beta naught for this research. An experiment was conducted using one radar scene to determine the suitability of $\sigma°$ or $\beta°$ for soil moisture estimation in the study site. It was found that backscatter values as $\beta°$ show better correlation with soil moisture than backscatter values as $\sigma°$ in the study site. It is believed that $\beta°$ produces a better estimation of soil moisture due to the low relief in the study site [27,28].

Removing Speckles

In radar remote sensing, coherent imaging systems produce images with a granular appearance. Bright and dark spots caused by random constructive and destructive interference of the wavelets returning from the various scatterers within the resolution cell of the system [29]. The effect of this interference process is regarded as a multiplicative noise, called speckle. Removing speckles is an essential component of SAR image pre-processing techniques. Speckles should be reduced, preserving edges and image texture to extract information from SAR imagery [30].

The best approach for removing speckles in SAR imagery is to use speckle removing filters. These speckle removing filters are developed on the basis of the multiplicative model with an exponential probability density function [31]. To remove the speckles in the acquired radar scenes different types of filtering techniques available in the Erdas Imagine image processing software were tested with different window sizes. The "Lee" filtering technique with 5 × 5 window size was found efficient in removing the speckles and was used to de-speckle the acquired radar scenes.

Geometric Correction

All SAR data were obtained as georectified using the Universal Transverse Mercator (UTM) projection system from the Alaska Satellite Facility (ASF). The GIS coverage of the sample locations and all other GIS data were georectified to the same projection system. A recent aerial photograph was used to aid the georectification process.

Digital Elevation Model (DEM)

The knowledge of the local incident angle is essential for the quantitative estimation of soil moisture and roughness from SAR data. In the absence of topographic relief the local incident angle equals the radar look angle. In areas with large topographic relief, the incident angle becomes a function of the radar look angle and the local terrain slope. This makes the straightforward surface parameter estimation difficult. It is then necessary to correct the SAR data for terrain to allow geometric overlays of remotely sensed data from different sensors and/or geometries. The elevation of the study site varies from 900 m to 1100 m. Despite the minimal topographic influence, a terrain correction was performed using USGS 30 m Digital Elevation Model (DEM) (Figure 3) to improve geometric accuracy of the data. Since topographic variations influence the distribution of soil moisture in the field the acquired DEM was also used in the SAR soil moisture estimation models.

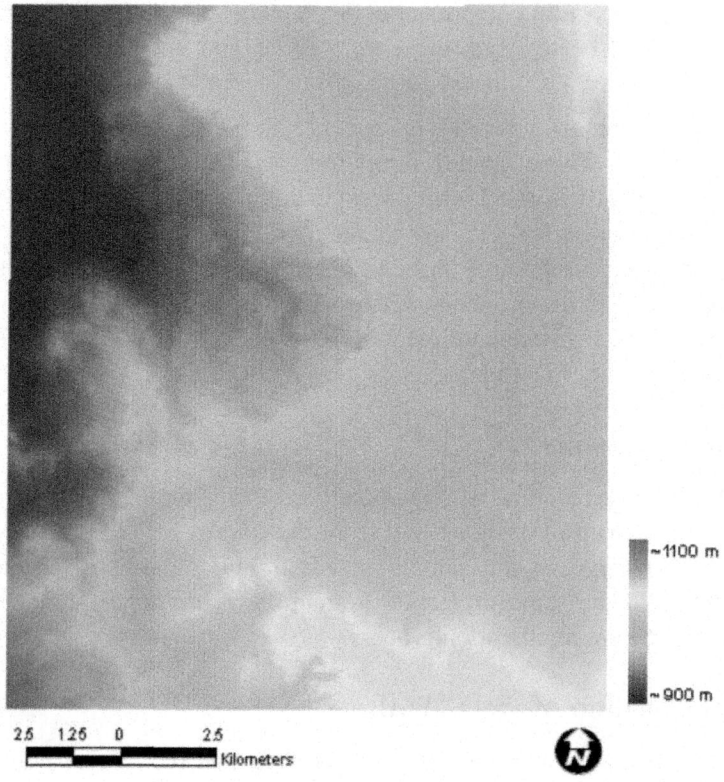

Figure 3: Digital Elevation Model (DEM) for the study site.

In Situ Soil Moisture Measurements

Near-real time soil moisture data is needed for quantitative estimation of soil moisture using SAR data. Soil samples were collected in the selected parts of Nash Draw and analyzed them to calculate volumetric soil moisture for calibrating the SAR imagery acquired on 2 August 2006 and 6 November 2006. Eighty soil samples were collected in the study site for each date.

Sampling Technique

A stratified soil sampling technique [32] was used to acquire the soil samples. The study site was divided into four equal parts and random sample points were selected in each part using 500 m grid spacing, with 20 samples collected in each part. Accessibility and variation in soil types was used to select sampling sites. Figure 4 shows the distribution of the soil sample locations in the study site.

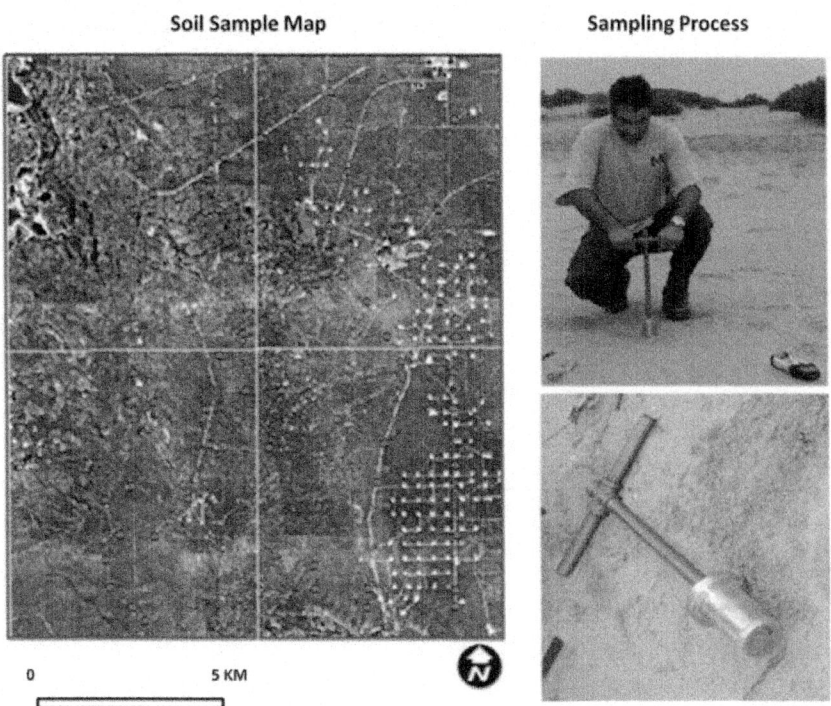

Figure 4: Distribution of soil samples collected for *in situ* soil moisture measurements.

RADARSAT 1 acquires imagery in C-band with HH polarization at 5.7 cm wavelength and 5.3 GHz frequency. According to [33] the RADARSAT 1 beam should be able to penetrate ground up to top few cm in dry bare soil condition. Therefore, samples were collected at a depth of approximately 3 cm below the land surface.

Volumetric Soil Moisture Measurement

Gravimetric soil moisture values were calculated from the obtained soil samples using the ASTM standard test method [34]. It was then converted to volumetric soil moisture using sample volume (Equations (5) and (6)). Table 1 shows the statistics of the soil moisture measurements for both August and November data sets.

$$V_w = \frac{(M_w - M_d)}{\rho_w}(cm^3) \tag{5}$$

$$\theta_v = \frac{V_w}{V_s} \times 100(\%) \tag{6}$$

where, V_w—the volume of water content in the obtained soil sample (cm³), M_s—the mass of moist soil (g), M_d—the mass of dry soil (g), ρ_w—water density (g/cm³), θ_v—volumetric soil moisture (%), and the volume of the sample container $V_s = 76.55$ (cm³).

Table 1: Statistics of *in situ* soil moisture measurements

Statistical Parameters	Measurements (in %)	
	1–3 August 2006	4–6 November 2006
Mean	6.48	2.80
Minimum	2.55	0.26
Maximum	14.53	10.16
Standard Deviation	2.56	2.13

Vegetation Data

Vegetation cover strongly influences soil moisture estimation due to the interaction of the microwaves with the vegetation and soil. The amount of vegetation, its dielectric properties, and distribution pattern can significantly affect the sensitivity of microwave backscatter to soil moisture. The SAR imagery time series data were used to produce vegetation distribution maps for the study site and a vegetation density map for each SAR image acquisition date.

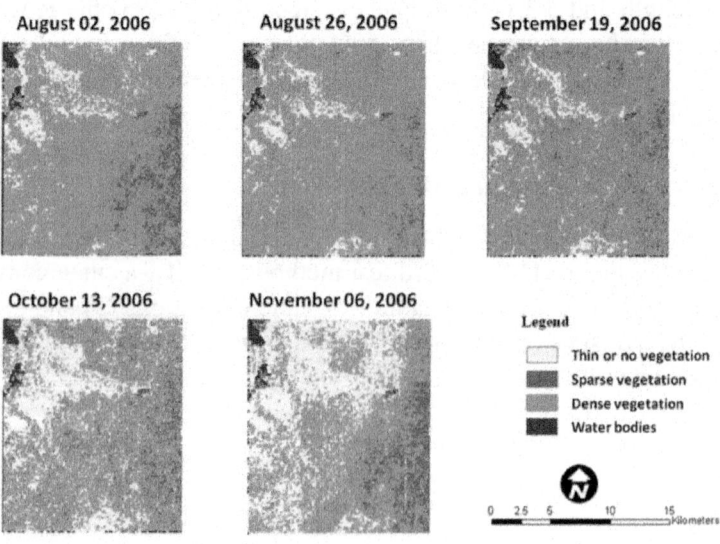

Figure 5: Vegetation density map for each SAR image acquisition date.

The ISODATA clustering algorithm was used to classify each SAR image to produce the vegetation density map for the study site. The image was classified into 25 classes and then recoded the data into four classes: thin or non-vegetated areas, sparsely vegetated areas, densely vegetated areas, and water. Field observations of vegetation were used to recode the data and produced vegetation density maps (Figure 5).

Soil Type and Soil Salinity

The soil data were acquired from the U. S. Department of Agriculture Natural Resource Conservation Service (USDA-NRCS) and converted into a 10 m gridded soil type dataset. This dataset was incorporated into the soil moisture estimation model. Figure 6 shows the soil type map for the study site. The soil type data was used in both multiple linear regressions and artificial neural networks based soil moisture estimation models.

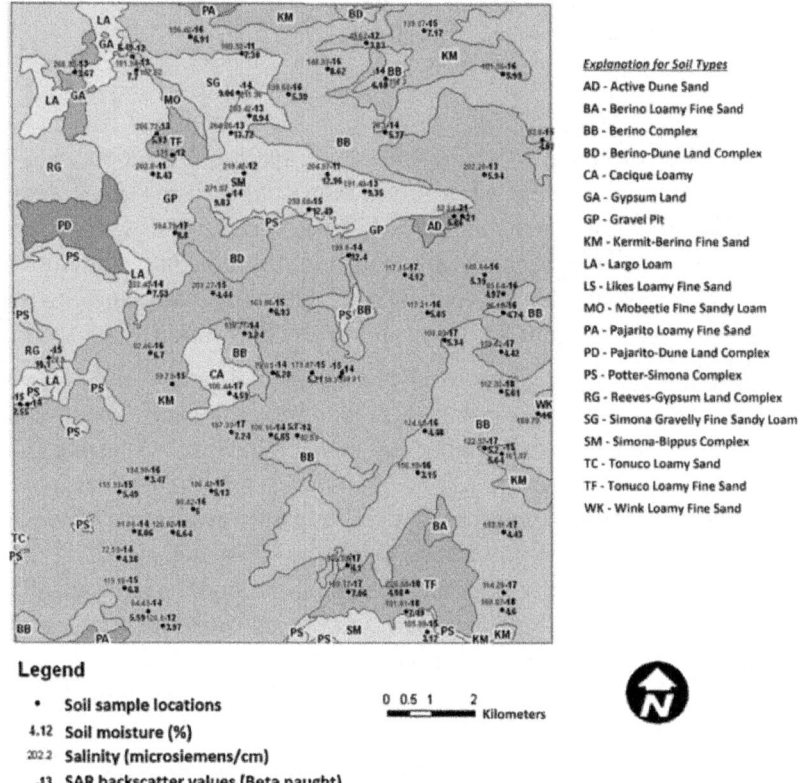

Explanation for Soil Types
AD - Active Dune Sand
BA - Berino Loamy Fine Sand
BB - Berino Complex
BD - Berino-Dune Land Complex
CA - Cacique Loamy
GA - Gypsum Land
GP - Gravel Pit
KM - Kermit-Berino Fine Sand
LA - Largo Loam
LS - Likes Loamy Fine Sand
MO - Mobeetie Fine Sandy Loam
PA - Pajarito Loamy Fine Sand
PD - Pajarito-Dune Land Complex
PS - Potter-Simona Complex
RG - Reeves-Gypsum Land Complex
SG - Simona Gravelly Fine Sandy Loam
SM - Simona-Bippus Complex
TC - Tonuco Loamy Sand
TF - Tonuco Loamy Fine Sand
WK - Wink Loamy Fine Sand

Legend

• Soil sample locations
4.12 Soil moisture (%)
202.2 Salinity (microsiemens/cm)
-13 SAR backscatter values (Beta naught)

0 0.5 1   2
                 Kilometers

Figure 6: Measurements of soil moisture, soil salinity, and radar backscatter values shown on different soil types in Nash Draw (August 2006 data set).

The salinity of the 80 soil samples (collected for *in situ* soil moisture measurements in August 2006) were measured using USDA Soil Water Extract 1:5 methods [35]. The measurement is expressed by Electrical Conductivity (EC) in µS/cm. The soil salinity measurements were also used in both multiple linear regressions and artificial neural networks based soil moisture estimation models. Figure 6 shows the distribution of measured soil moisture, soil salinity, and radar backscatter values of different soil types in Nash Draw, NM.

METHODS

In this study, multiple linear regressions and Artificial Neural Networks (ANN) based models were used to investigate the influence of vegetation, elevation, soil type, and salinity on soil moisture estimation using microwave imagery. Coefficient of determination (R^2) values were used to evaluate the suitability of the different numerical models to estimate and map soil moisture distributions. Soil moisture maps were prepared for the five SAR imagery acquisition dates using the most suitable numerical model.

Multiple Linear Regressions

In semi-arid environments, the influence of sparse vegetation was found negligible in several soil moisture estimation studies using SAR data, e.g., [11,16,17,18,19]. However, as shown by [23] the vegetation distribution pattern in Nash Draw could significantly influence the soil moisture estimation using high resolution SAR data. The obtained vegetation density and distribution maps of Nash Draw (Figure 5) also support this observation.

In estimating soil moisture from low frequency radar backscattering (L band radar), the di-electric constant was found to be weakly sensitive to soil types [12]. The sensitivity of higher frequency radar backscattering (C band radar) to soil types, however, has not been fully analyzed in the published literature. Several studies reported that soil moisture estimation using L-band radar should consider the impact of soil salinity in the model [36,37,38]. Since in this research the soil moisture estimation model used C band radar, the impact of both soil type and soil salinity on soil moisture estimation was investigated. Multiple linear regressions were conducted incorporating vegetation density information, elevation, soil type, and soil salinity in addition to radar backscattering (measured as $\beta°$) to estimate soil moisture in the study site. The regression was done in a step fashion where independent variables (vegetation density information, elevation, soil type, and soil salinity) were sequentially added to the regression analysis to evaluate the effects of each independent variable in soil moisture estimation.

Equations (7)–(11) are the numerical models developed by multiple linear regressions with the observed *in situ* soil moisture measurements and different combination of independent variables to estimate soil moisture in Nash Draw, NM. The observed soil moisture data were acquired from the 2 August data set. Table 2 shows the coefficient of determination (R^2) values of the corresponding models. From the results of the simple linear regressions it was found that both the 2 August and the 6 November data sets produced similar results. Therefore, it was decided to use only one data set to evaluate the model performance.

$$\theta_w = 16.01 + 0.50\beta^o - 0.86V \tag{7}$$

$$\theta_w = 13.90 + 0.42\beta^o - 0.79V + 0.22S \tag{8}$$

$$\theta_w = 26.85 + 0.40\beta^o - 0.81V + 0.01E \tag{9}$$

$$\theta_w = 18.17 + 0.38\beta^o - 0.77V + 0.21S - 0.01E \tag{10}$$

$$\theta_w = 24.91 + 0.27\beta^o - 0.97V + 0.14S - 0.02E - 0.01SL \tag{11}$$

where, β^o—Backscatter value, V—Vegetation, S—Soil type, E—Elevation, and SL—Soil salinity.

Table 2: Coefficient of determination (R^2) values of multiple linear regression based models.

Independent Variables	Coefficient of Determination (R^2)	Insignificant Variables
β^o, V	0.50	--
β^o, V, S	0.60	--
β^o, V, E	0.51	E
β^o, V, S, E	0.61	E
β^o, V, S, E and SL	0.66	E and SL

Note: β^o—Backscatter value; V—Vegetation; S—Soil type; E—Elevation; and SL—Soil salinity.

Artificial Neural Networks (ANN)

Most SAR based soil moisture estimation models are based on the assumption that soil moisture distribution is linearly related to the radar backscatter of the soil moisture surface, e.g., [11,12,21,39]. There are a few studies that explore the non-linear relationships between soil surface moisture and radar backscatters. In this study, Artificial Neural Networks (ANN) based numerical models were developed to estimate soil surface moisture from SAR data and to explore the non-linear relationship between soil moisture and SAR backscatters. The significance of vegetation coverage, elevation, soil type, and soil salinity in soil moisture estimation using SAR data and artificial neural networks based models was also investigated.

Artificial neural networks, are a branch of artificial intelligence [40] in which the solution to a problem is learned from a set of examples [41]. A neural network can be regarded as a nonlinear mathematical function, which transforms a set of input variables into a set of output variables. The use of neural networks has been shown to be effective alternatives to more traditional statistical techniques [42]. Neural networks can be trained to approximate any smooth, measurable function [42], can model highly non-linear functions, and can be trained to be accurately generalized when presented with unseen data [26]. In a typical neural network model, a single neuron forms a weighted sum of the inputs $x_1, x_2, ..., x_d$ given by $a = {}_iw_ix_i$ and then transforms this sum using a non-linear activation function $g()$ to give a final output $z = g(a)$ (Figure 7).

A feed forward neural network can be regarded as a nonlinear mathematical function, which transforms a set of input variables into a set of output variables. The multilayer perceptron is the most widely used feed forward neural network. Figure 7 shows a single processing unit of neural networks. If we consider a set of m such units, all with common inputs, then we arrive at a neural network having a single layer of adaptive parameters (weights) as illustrated in Figure 8. The output variables are denoted by z_j and are given by Equation (12).

$$z_j = g\left(\sum_{i=0}^{d} w_{ji}x_i\right)$$

$$(12)$$

where w_{ji}—the weight from input i to unit j, and $g()$—an activation function as discussed previously.

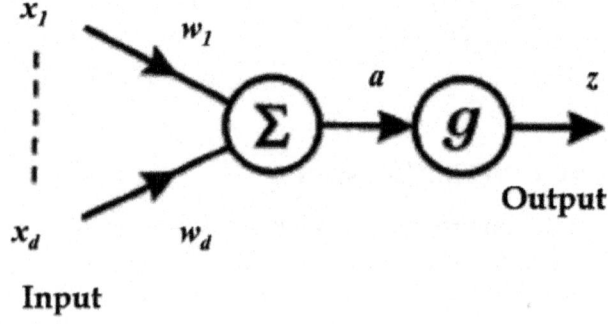

Figure 7: A single processing unit in neural networks.

Artificial neural networks, based non-linear numerical models were developed for soil moisture estimation for the entire study site using the 2 August data set. Five different neural networks based models were developed, through addition of different variables, to estimate soil moisture in Nash Draw. The first model had only one input, the backscatter values. This model uses the non-linear relationship between radar backscatter and soil moisture content. The other four models used additional inputs (e.g., vegetation coverage, elevation, soil type, and soil salinity) in different combinations with radar backscatter values. JMP statistical software was used to perform the neural networks based analysis. The model coefficient of determination (R^2) and cross validation coefficient of determination (CV R^2) values were used to evaluate the model performance for soil moisture prediction. The impact of soil salinity was investigated and inclusion of this variable did not significantly improve model performance. Figure 9 shows the simplified schematic of the models that were developed. Table 3 shows the R^2 and CV R^2 values of the corresponding models.

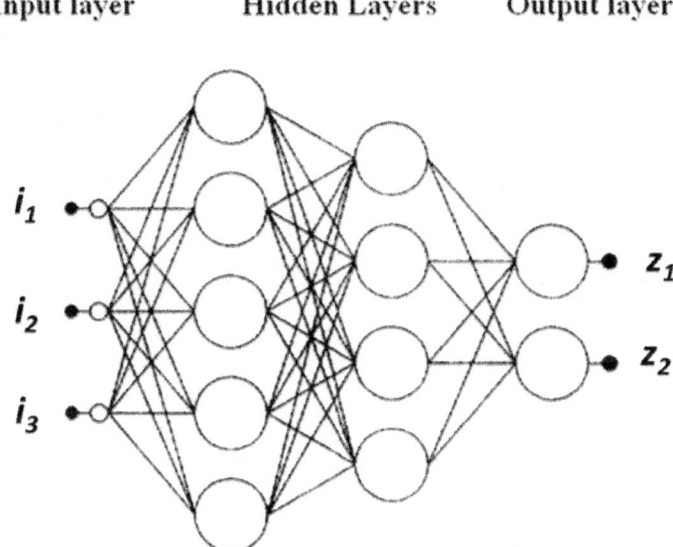

Figure 8: A multilayer perceptron with two hidden layers.

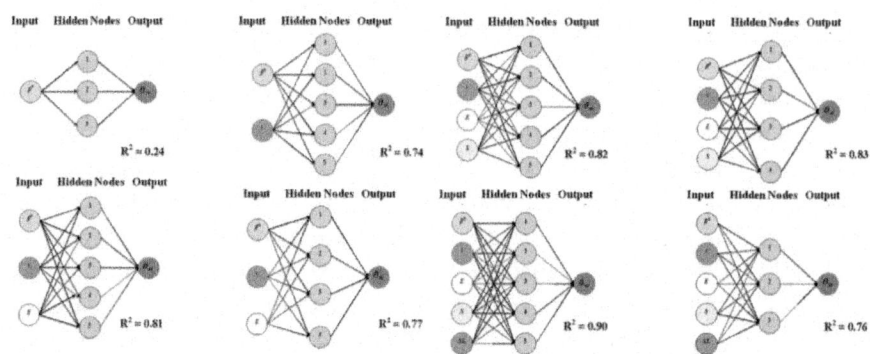

Figure 9: Neural networks based numerical models for soil moisture estimation in Nash Draw, NM, USA. Note: °—Backscatter value; *V*—Vegetation; *S*—Soil type; *E*—Elevation; *SL*—Soil salinity.

Table 3: Coefficient of determination (R^2) and cross validation coefficient of determination (CV R^2) of the neural networks based numerical models for soil moisture estimation.

Input Variables	Hidden Nodes	Coefficient of Determination (R^2)	Cross Validation (CV) R^2
$\beta°$	3	0.24	0.11
$\beta°, V$	5	0.74	0.49
$\beta°, V, E$	4	0.77	0.54
$\beta°, V, E$	5	0.81	0.44
$\beta°, V, E, S$	4	0.83	0.56
$\beta°, V, E, S$	5	0.82	0.47
$\beta°, V, E, S$ and SL	3	0.76	0.45
$\beta°, V, E, S$ and SL	4	0.82	0.54
$\beta°, V, E, S$ and SL	5	0.90	0.58

Note: $\beta°$—Backscatter value; *V*—Vegetation; *S*—Soil type; *E*—Elevation; *SL*—Soil salinity.

RESULTS

Soil Moisture Estimation

Near-real time field observations, acquired at the beginning and end of the time series, in conjunction with SAR data were used to estimate soil moisture from the SAR imagery acquired on five different dates in 2006. Model coefficient of determination (R^2) and model cross validation coefficient of determination (CV R^2) (for non-linear models) values were compared and used to evaluate their suitability for soil moisture estimation in Nash Draw. The model with the highest R^2 and CV R^2 values was considered as the most appropriate model for soil moisture estimation. The accuracy of the selected models was evaluated by Kappa statistics. The selected soil moisture estimation numerical models were then used to convert *β°* SAR data into soil moisture data. The soil moisture data was divided into several categories to aid in the interpretation of the spatial distribution of the soil moisture in the study site.

The following observations were made after the evaluation of the R^2 and CV R^2 values of the models developed for soil moisture estimation in Nash Draw, NM.

Simple linear regression between radar backscatter values and *in situ* soil moisture measurements can be used to develop SAR-based soil moisture estimation models with model R^2 values of 0.51 to 0.61, but application of the model should be restricted to non-vegetated to thinly vegetated areas [23].

Multiple linear regressions using radar backscatter values, vegetation density, soil type, and elevation as independent variables can be used to develop soil moisture estimation models for the entire study site, including areas with thicker vegetation.

Neural networks based models, using radar backscatter values, vegetation density, soil type, and elevation can also be used to estimate soil moisture for the entire study site, including areas with thicker vegetation. Neural networks based models achieved higher R^2 values and performed better than multiple linear regressions based models.

A neural network based numerical model using radar backscatter values, vegetation density, soil type and elevation was used to estimate soil moisture in the study site. This model was developed for both the 2 August and the 6 November data sets. The R^2 and CV R^2 values for the August model were 0.83 and 0.56 respectively, and 0.81 and 0.55 for the November model, respectively.

The model developed for the 2 August data set was also used to map soil moisture for the 26 August data set. The model developed for the 6 November data set was also used to map soil moisture for the 13 October data set. Since 19 September is approximately temporally equal from both 2 August and 6 November, we applied both the 2 August and the 6 November models to the 19 September data set and estimated soil moisture by taking the average of the two estimations.

Two sets of 50 m resolution soil moisture data were produced for each of the five dates of SAR data for the study site. The first set includes the unclassified soil moisture data, where the value of each pixel of the dataset is the volumetric soil moisture estimation (Figure 10). In the second dataset each pixel is classified into six categories of soil moisture to map the spatial variations in soil moisture at 2.5% intervals (Figure 11).

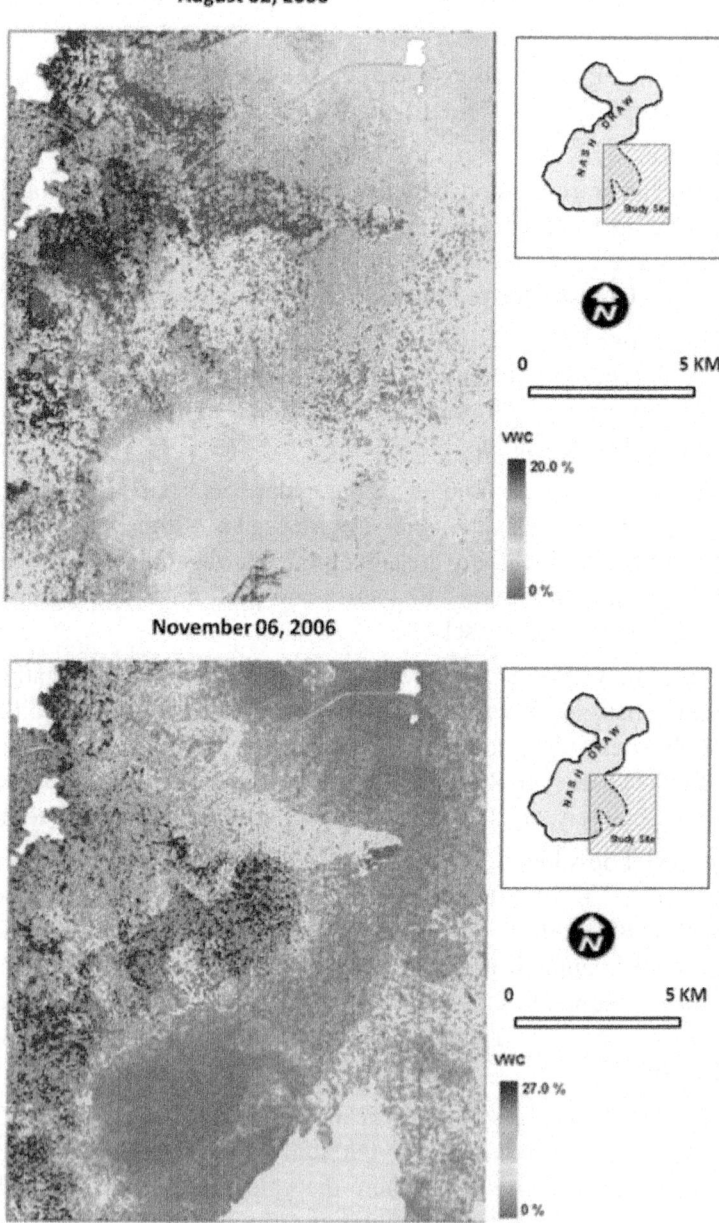

Figure 10: Unclassified soil moisture data generated for the study site.

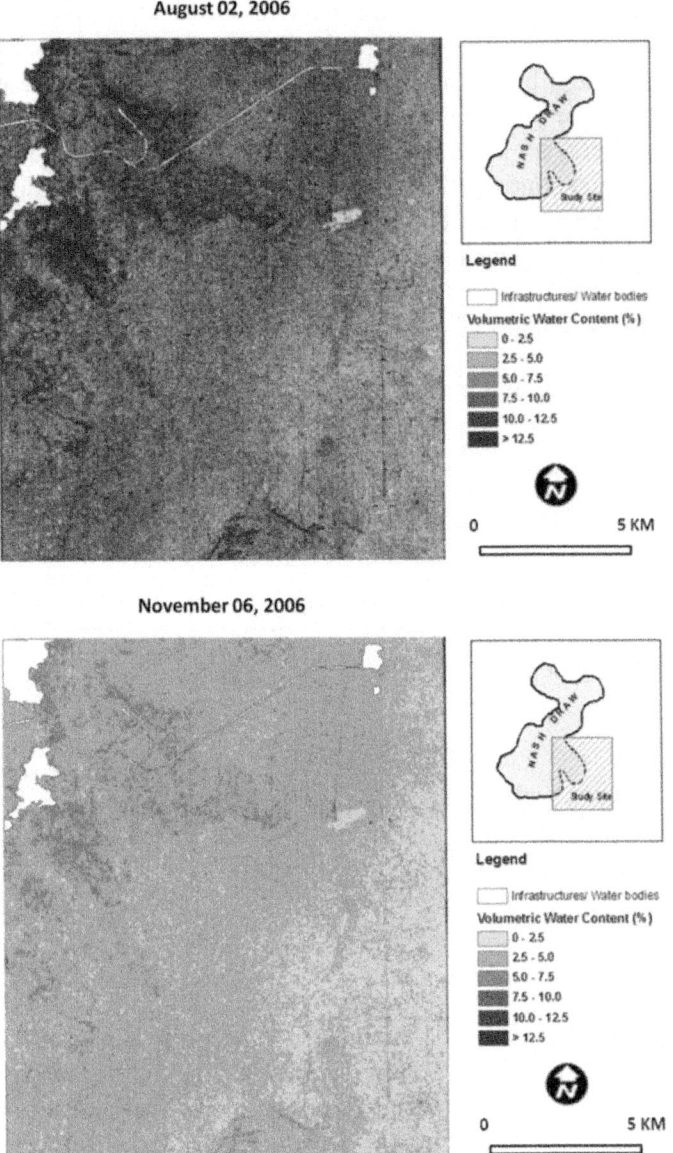

Figure 11: Classified soil moisture data generated for the study site.

Accuracy Assessment

Kappa statistics [43,44] were used to evaluate the accuracy of the soil moisture estimation data produced from the SAR imagery. Kappa statistics have been used successfully for accuracy assessment in different remote sensing based studies, e.g., [13,45,46,47]. It is a discrete multivariate technique of accuracy assessment [28]. Kappa coefficients express the proportionate reduction in error generated by a classification process compared with the error of a completely random classification. For example, a value of 0.82 implies that the classification process is avoiding 82% of the errors that a completely random classification generates [43]. Kappa can be thought of as the chance-corrected proportional agreement, and possible values range from +1 (perfect agreement) to −1 (complete disagreement). A value of 0 indicates no agreement above that expected by chance. The calculation of the Kappa coefficients is explained below using an example of a 2 × 2 matrix (Table 4).

Table 4: Computation of Kappa coefficients

–		SAR Soil Moisture		Total
		Class 1	Class 2	
Reference Soil Moisture	Class 1	P_{11}	P_{12}	$P_{11} + P_{12} = \alpha$
	Class 2	P_{21}	P_{22}	$P_{21} + P_{22} = \beta$
Total		$P_{11} + P_{21} = \gamma$	$P_{12} + P_{22} = \delta$	$P_{11} + P_{22} = \chi$
		Total number of data points: $Q = \alpha + \beta = \gamma + \delta$		

Observed agreement:

$$P_o = \frac{\chi}{Q} \tag{13}$$

Chance agreement:

$$P_e = \frac{\alpha}{Q} \times \frac{\gamma}{Q} + \frac{\delta}{Q} \times \frac{\beta}{Q} \tag{14}$$

Kappa coefficient:

$$K = \frac{P_o - P_e}{1 - P_e} \tag{15}$$

Kappa analysis requires continuous ground truth data so that a sufficient amount of random reference data can be obtained. Therefore, a continuous soil moisture surface was created from the *in situ* soil moisture data, using kriging [48]. Jackknife resampling techniques was used to correct for bias [49]. The measured *in situ* soil moisture data were randomly divided into a training set (90%) and testing set (10%) and were analyzed using the Arc GIS Geostatistical Analyst software. The training data set was used to create the krigged soil moisture surface and the testing data was used to evaluate the kriging results. Ten different krigged soil moisture data sets were generated using 10 different training data sets obtained from the same *in situ* soil moisture measurements. The RMS error, average standard error, standardized mean of error, and standardized RMS error (obtained from the evaluation of kriging results with the testing data) were used to select the appropriate kriging surface.Figure 12 shows the krigged surface generated for the 2 August and 6 November data sets.

Both SAR derived soil moisture data and the krigged soil moisture prediction surfaces were categorized into three class intervals. For the 2 August data set, the classes were 0.0%–5.0%, 5.0%–10.0%, and >10.0%. Since the soil moisture values were much lower in the November data set, these ranges were 0.0%–2.5%, 2.5%–5.0%, and >5.0%. Three hundred randomly generated points were used to calculate the Kappa coefficients and perform an accuracy assessment. Erdas Imagine image processing software was used to perform this accuracy assessment and Kappa analysis. Kappa coefficients were calculated for both the individual classes and the whole data sets.

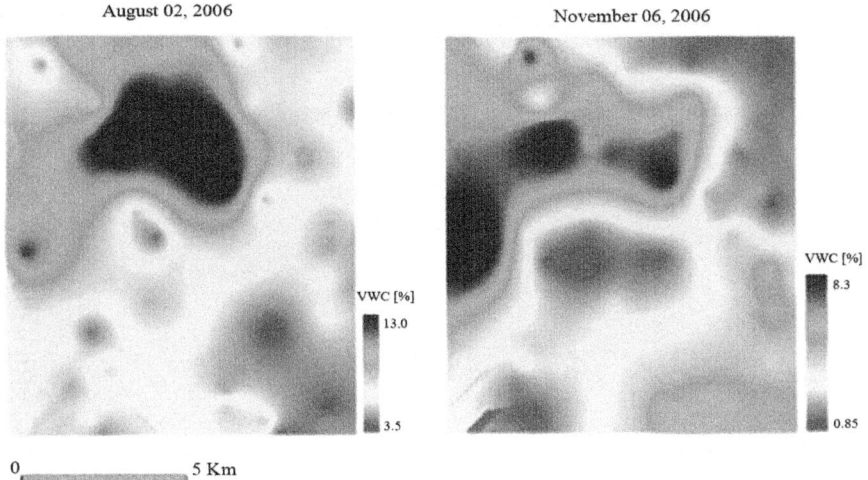

Figure 12. Krigged soil moisture surface generation for 2 August and 6 November 2006 field data field data.

The overall Kappa coefficients obtained for the 2 August and the 6 November data sets were 0.43 and 0.61, respectively. The overall accuracy was 75.67% and 77.67% for the 2 August and the 6 November data sets, respectively. Figure 13 and Figure 14 show the Kappa coefficients and classification accuracy for the individual classes. The evaluation of the individual classes indicate that the developed model performed well at low soil moisture regimes compared to the high soil moisture regimes in both 2 August and 6 November data sets.

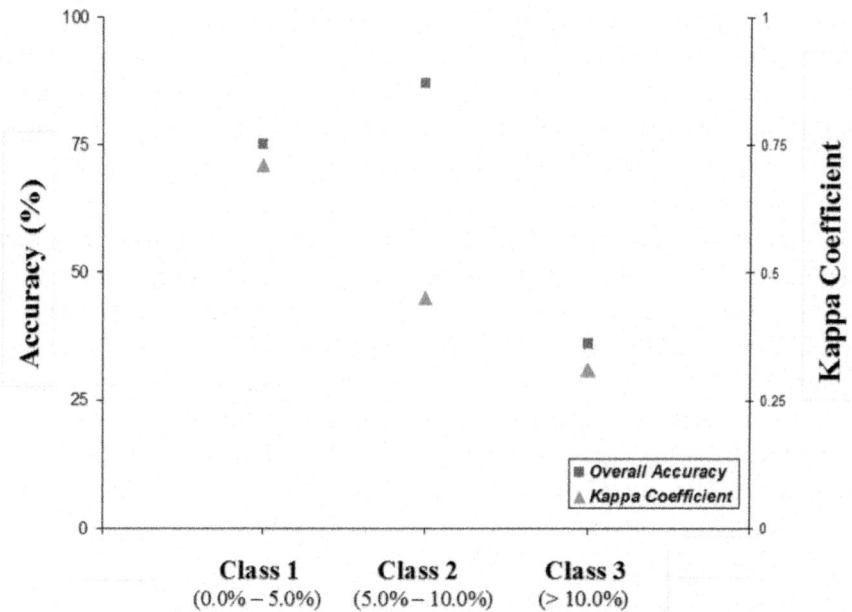

Figure 13: Kappa statistics for 2 August 2006 data set.

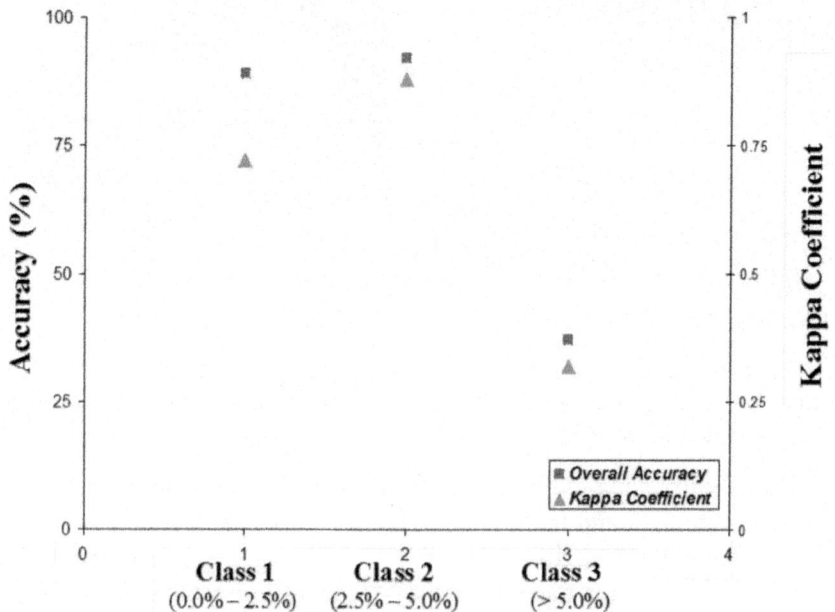

Figure 14: Kappa statistics for 6 November 2006 data set.

DISCUSSIONS

Research conducted by [23] has shown that in semi-arid environments even sparse vegetation can adversely influence the accuracy of soil moisture values estimated from radar backscatter (obtained from high resolution SAR data) using a linear relationship. This observation is supported by lower R^2 values (0.24 and 0.05 for the August and the November datasets respectively) for the linear numerical models developed for the entire study site and higher R^2 values (0.61 and 0.52 for the August and the November datasets respectively) for linear numerical models developed for portions of the study site with little or no vegetation. The lowest R^2 values were from linear numerical models developed for the more densely vegetated portions of the study site (0.01 to 0.04).

This research shows that soil moisture estimation using high resolution SAR data in a semi-arid environment can be improved by developing numerical models that use multiple linear regressions incorporating additional variables such as, vegetation density, soil type, and elevation in addition to radar backscatter values. Regression coefficients as high as 0.61 were obtained for a numerical model covering the entire study site using multiple linear regressions.

The non-linear relationship between radar backscatter values and soil moisture was also investigated by [23] using artificial neural networks. It showed that a neural network developed using only radar backscatter values and soil moisture had an R^2 value of 0.24, similar to the R^2 value obtained from numerical models using simple linear regressions. This study obtained significant improvement in soil moisture estimation when additional variables such as, vegetation density, soil type, and elevation were added as inputs to the artificial neural networks based models. The use of these additional inputs results in coefficients of determination of 0.83 and 0.81 for the entire study site for the 2 August and the 6 November datasets, respectively. The cross validation coefficients of determination (CV R^2) of the same models were 0.56 and 0.55, respectively.

Soil moisture data were produced using artificial neural networks based non-linear models that incorporated inputs of vegetation density, soil type, elevation, and radar backscatter values. The accuracy of the modeled soil moisture data was also evaluated by the comparison of a soil moisture distribution surface from the models with a soil moisture surface obtained by kriging the *in situ* measurements. The comparison was done by Kappa statistics. The overall accuracy and the Kappa coefficient for the soil moisture data obtained for 2 August and 6 November were 75.67% and 77.67%, and 0.43 and 0.61, respectively. Although the obtained kappa coefficient values indicate overall good agreement between the measured and estimated soil moisture values, however, it is worth looking into the possibility of having the influence of autocorrelation between these two soil moisture data sets. As suggested by [50] since most thematic maps have some degrees of spatial autocorrelation, the calculated expected agreement (by Kappa statistics) could be usually higher than the true expected agreement.

CONCLUSIONS AND FUTURE WORK

This research demonstrates the proof of concept of the application of Artificial Neural Networks (ANN) based models for estimating soil surface moisture in semi-arid environments of south-eastern New Mexico using high resolution SAR data (Radarsat 1 SAR Fine data). It is strongly believed that the methods developed in this research can be used to produce soil moisture data from high resolution SAR imagery in semi-arid environments, such as Nash Draw in New Mexico. The hydrology of Nash Draw in New Mexico is characterized by evaporite karst and is not well understood. The data produced in this research should be very useful to identify the pattern of soil moisture distribution in space and time and contribute to a better understanding of the groundwater recharge in the study area.

In the future, to develop an operational soil moisture estimation tool (for this particular area) using high resolution SAR data, the proposed modeling effort can be further enhanced/improved by considering: (1) inclusion of surface roughness parameters in the artificial neural networks based models in addition to vegetation density, soil type, and elevation; (2) incorporation of near real-time high resolution optical satellite data for mapping vegetation density; and (3) involving other statistical tests such as Root Mean Squared Error (RMSE), Mean Absolute Percent Error (MAPE) *etc.* for model accuracy assessments. It is also recommended to use high resolution L-band SAR data with the capability of multi-polarization. L band data will provide the opportunity to obtain moisture estimation in greater depth.

ACKNOWLEDGMENTS

Authors thankfully acknowledge the NASA Applied Sciences Program and the University of Mississippi Geoinformatics Center (UMGC) for funding this study through a Rapid Prototyping Capability (RPC) for Earth-Sun Systems Sciences project at The University of Mississippi. Thanks are also due to the Alaska Satellite Facility (ASF) for providing the SAR imagery; Robert Holt for proving technical support for regression and artificial neural networks analysis; Dennis Powers, Glen Garrett, and Dirk O'Daniel for their assistance in the field to acquire soil samples for soil moisture analysis; and Patrick

Yamnik for assisting in image processing and GIS analysis. Authors are also grateful to Chung R. Song for his continuous support throughout the study.

AUTHOR CONTRIBUTIONS

Azad Hossain is responsible for data acquisition, processing and analysis (soil moisture estimation and accuracy assessments. Greg Easson is responsible for digital image processing aspect. Both authors prepare this manuscript together.

REFERENCES

1.   Hossain, F.; Anagnostou, E.N. Numerical investigation of the impact of uncertainties in satellite rainfall and land surface parameters on simulation of soil moisture. Adv. Water Resour. 2005,28, 1336–1350.

2.   Georgakakos, K.P.; Baumer, O.W. Measurement and utilization of on-site soil moisture data. J. Hydrol. 1996, 184, 131–152.

3.   Lakshmi, V. Remote sensing of soil moisture. ISRN Soil Sci. 2013, 2013, 1–33.

4.   Robock, A.; Luo, L.; Wood, E.F.; Wen, F.; Mitchell, K.E. Evaluation of the North American Land Data Assimilation System over the southern Great Plains during the warm season. J. Geophys. Res. 2003, 108.

5.   Zhang, L.; Dawes, W.R.; Hatton, T.J.; Reece, P.H.; Beale, G.T.H.; Packer, I. Estimation of soil moisture and groundwater recharge using the TOPOG_IRM model. Water Resour. Res. 1999, 35, 149–161.

6.   Silva, C.S.D.; Rushton, K.R. Groundwater recharge estimation using improved soil moisture balance methodology for a tropical climate with distinct dry seasons. Hydrol. Sci. J. 2007, 52, 1051–1067.

7.   Hossain, A.; Easson, G. Mapping spatial variation in surface moisture using reflective and thermal ASTER imagery for Southern Africa. In Proceedings of 2006 ASPRS Annual Conference, Reno, NV, USA, 1–5 May 2006.

8.   Narasimhan, B.; Srinivasan, R.; Arnold, J.G.; di Luzio, M. Estimation of long-term soil moisture using a distributed parameter hydrologic model and verification using remotely sensed data.Trans. ASAE 2005, 48, 1101–1113.

9.   Albergel, C.; de Rosnay, P.; Gruhier, C.; Muñoz-Sabater, J.; Hasenauer, S.; Isaksen, L.; Kerr, Y.; Wagner, W. Evaluation of remotely sensed and modeled soil moisture products using global ground-based in-situ observations. Remote Sens. Environ. 2012, 118, 215–226.

10.   Klemas, V.; Finkl, C.W.; Kabbara, N. Remote sensing of soil moisture:

An overview in relation to coastal soils. J. Coast. Res. 2014, 30, 685–696.

11. Dubois, P.C.; van Zyl, J.; Engman, E.T. Measuring soil moisture with imaging radars. IEEE Trans. Geosci. Remote Sens. 1995, 33, 915–926.

12. Ulaby, F.T.; Dubois, P.C.; van Zyl, J. Radar mapping of surface soil moisture. J. Hydrol. 1996,184, 57–84.

13. Jackson, T.J. Measuring surface soil moisture using passive microwave remote sensing. Hydrol. Process. 1993, 7, 139–152.

14. Jackson, T.J.; Schmugge, J.; Engman, E.T. Remote sensing applications to hydrology: Soil moisture. Hydrol. Sci. J. 1996, 41, 517–530.

15. Borgeaud, M.; Saich, P. Status of the retrieval of bio- and geophysical parameters from SAR data for land applications. In Proceedings of the 1999 IEEE International Geoscience and Remote Sensing Symposium, Hamburg, Germany, 28 June–2 July 1999.

16. Chanzy, A.; Kerr, Y.; Wigneron, J.P.; Calvet, J.C. Soil moisture estimation under sparse vegetation using microwave radiometry at C-band. In Proceedings of the 1997 International Geoscience and Remote Sensing Symposium, Singapore, 3–8 August 1997.

17. Demircan, A.; Rombach, M.; Mauser, W. Extraction of soil moisture from multitemporal ERS-1 SLC data of the Freiburg test site. In Proceedings of the 1993 IEEE International Geoscience and Remote Sensing Symposium, Tokyo, Japan, 18–21 August 1993.

18. Lin, D.S.; Wood, E.F. Behavior of AirSAR signals during MAC-Europe'91. In Proceedings of the 1993 IEEE International Geoscience and Remote Sensing Symposium, Tokyo, Japan, 18–21 August 1993.

19. Thoma, D.P.; Moran, M.S.; Bryant, R.; Rahman, M.; Holifield-Collins, C.D.; Skirvin, S.; Sano, E.E.; Slocum, K. Comparison of four models to determine surface soil moisture from C-band radar imagery in a sparsely vegetated semiarid landscape. Water Resour. Res. 2006, 42, 1–12.

20. Mladenovaa, I.E.; Jacksona, T.J.; Njokub, E.; Bindlishc, R.; Chanb, S.; Cosha, M.H.; Holmesa, T.R.H.; de Jeud, R.A.M.; Jonese, L.; Kimballe, J.; et al. Remote monitoring of soil moisture using passive microwave-based techniques—Theoretical basis and overview of selected algorithms for AMSR-E. Remote Sens. Environ. 2014, 144, 197–213.

21. Moran, M.S.; Hymer, D.C.; Qi, J.; Sano, E.E. Soil moisture evaluation using multi-temporal Synthetic Aperture Radar (SAR) in semiarid rangeland. Agric. For. Meteorol. 2000, 105, 69–80.

22. Moran, M.S.; Peters-Lidard, C.D.; Watts, J.M.; McElroy, S. Estimating soil moisture at the watershed scale with satellite-based radar and land

surface models. Can. J. Remote Sens. 2004,30, 805–826.

23. Hossain, A.; Easson, G. Microwave remote sensing of soil moisture. In Geoscience and Remote Sensing; Ho, P.P., Ed.; IN-TECH Press: Vukovar, Croatia, 2009; pp. 529–554.

24. Holt, R.M.; Beauheim, R.L.; Powers, D.W. Predicting fractured zones in the Culebra Dolomite. InDynamics of Fluids and Transport in Fractured Rock: AGU Geophysical Monograph Series; Faybishenko, B., Witherspoon, P.A., Gale, J., Eds.; American Geophysical Union: Washington, DC, USA, 2005; pp. 103–116.

25. Powers, D.W.; Beauheim, R.L.; Holt, R.M.; Hughes, D.L. Evaporite karst features and processes at Nash Draw, Eddy County, New Mexico. In Caves & Karst of Southeastern New Mexico, NM Geological Society Fifty-seventh Annual Field Conference Guidebook; Land, L., Lueth, V.W., Raatz, W., Boston, P., Love, D.L., Eds.; New Mexico Geological Society: Socorro, NM, USA, 2006; pp. 253–266.

26. Glenn, N.F.; Carr, J.R. Establishing a relationship between soil moisture and RADARSAT-1 SAR data obtained over the Great Basin, Nevada, USA. Can. J. Remote Sens. 2004, 30, 176–181.

27. Hossain, A.; Easson, G.; Powers, D.W.; Holt, R.M. 2007 Impact of Variation in Reflectivity of Microwave Data for Soil Moisture Estimation in Semi-arid Environment; Sigma Xi Poster Presentation; The University of Mississippi: University, MS, USA, 2007.

28. Hossain, A.; Easson, G.; Holt, R.M.; Powers, D.W. Developing methods for mapping soil moisture in Nash Draw, NM using Radarsat 1 SAR fine imagery. In Proceedings of the 2006 AGU Fall Meeting, San Francisco, CA, USA, 11–15 December 2006.

29. Goodman, J.W. Statistical properties of laser speckle patterns. In Laser Speckle and related Phenomena; Springe: Berlin, Germany, 1975; pp. 9–75.

30. D'Elia, C.; Ferraiuolo, G.; Pascazio, V.; Schirnzi, G. An MRF based technique for speckle reduction in SAR images. In Proceedings of the 2004 International Geoscience and Remote Sensing Symposium, Anchorage, AK, USA, 20–24 September 2004.

31. Touzi, R. A review of speckle filtering in the context of estimation, theory. IEEE Trans. Geosci. Remote Sens. 2002, 40, 2392–2404.

32. Dane, J.H.; Topp, G.C. Methods of Soil Analysis, Part 4, Physical Methods; Soil Science Society of America: Madison, WI, USA, 2002.

33. Boisvert, J.B.; Gwyn, Q.H.J.; Brisco, B.; Major, D.; Brown, R.J.

Evaluation of soil moisture techniques and microwave penetration depth for radar applications. Can. J. Remote Sens. 1995,21, 110–123.

34. ASTM D2216. Standard Test Method for Laboratory Determination of Water (Moisture) Content of Soil and Rock by Mass; American Society Testing and Materials: Philadelphia, PA, USA, 1999.

35. Tanji, K.K. Agricultural Salinity Assessment and Management; ASCE Manuals and Reports on Engineering Practice No. 71; American Society of Civil Engineers: Reston, VA, USA, 1990.

36. Shao, Y.; Guo, H.; Hu, Q.; Lu, Y.; Dong, Q.; Han, C. Effect of dielectric properties of moist salinized soils on backscattering coefficients extracted from RADARSAT image. IEEE Trans. Geosci. Remote Sens. 2003, 41, 1879–1888.

37. Sreenivas, K.; Venkataratnam, L.; Narasimha Rao, P.V. Dielectric properties of salt-affected soils. Int. J. Remote Sens. 1995, 16, 641–649.

38. Taylor, G.R.; Mah, A.H.; Kruse, F.A.; Kierein-Yong, K.S.; Hewson, R.D.; Bennette, B.A. Characterization of saline soils using airborne radar imagery. Remote Sens. Environ. 1996, 57, 127–142.

39. Dobson, M.C.; Pierce, L.; Arabandi, K.; Ulaby, F.T.; Sharik, T. Preliminary analysis of ERS-1 SAR for forest ecosystem studies. IEEE Trans. Geosci. Remote Sens. 1992, 30, 203–221.

40. Gardner, M.W.; Dorling, S.R. Artificial neural networks (The multilayer perceptron)—A review of applications in the atmospheric sciences. Atmos. Environ. 1998, 32, 2627–2636.

41. Bishop, C.M. Neural networks and their applications. Rev. Sci. Instrum. 1994, 65, 1803–1833.

42. Hornik, K.; Stinchcombe, M.; White, H. Multilayer feed forward networks are universal approximators. Neural Netw. 1989, 2, 259–366.

43. Cohen, J. A coefficient of agreement for nominal scales. Educ. Psychol. Meas. 1960, 20, 37–46.

44. Congalton, R. A review of assessing the accuracy of classifications of remotely sensed data.Remote Sens. Environ. 1991, 37, 35–46.

45. Congalton, R.G.; Green, K. Assessing the Accuracy of Remotely Sensed Data: Principles and Practices; Lewis Publisher: Boca Raton, FL, USA, 1999.

46. Naesset, E. Use of the weighted Kappa coefficient in classification error assessment of thematic maps. Int. J. Geogr. Inf. Syst. 1996, 10, 591–603.

47. Monserud, R.A.; Leemans, R. Comparing global vegetation maps with the Kappa statistic. Ecol. Model. 1992, 62, 275–293.

48. Isaaks, E.H.; Srivastava, R.M. An Introduction to Applied Geostatistics; Oxford University Press: Oxford, UK, 1989.

49. Efron, B.; Tibshirani, R. Bootstrap methods for standard errors, confidence intervals, and other measures of statistical accuracy. Stat. Sci. 1986, 62, 275–293.

50. Hagen-Zanker, A. An improved Fuzzy Kappa statistic that accounts for spatial Autocorrelation.Int. J. Geogr. Inf. Sci. 2009, 23, 61–73.

Chapter 4

ANTHRAX AND THE GEOCHEMISTRY OF SOILS IN THE CONTIGUOUS UNITED STATES

Dale W. Griffin [1], Erin E. Silvestri [2], Charlena Y. Bowling [2], Timothy Boe [3], David B. Smith [4] and Tonya L. Nichols [5]

[1]Coastal and Marine Science Center, U.S. Geological Survey, 600 4th Street South, St. Petersburg, FL 33701, USA

[2]National Homeland Security Research Center, U.S. Environmental Protection Agency, 26 W. Martin Luther King Drive, MS NG16, Cincinnati, OH 45268, USA

[3]National Homeland Security Research Center, Oak Ridge Institute for Science and Education, with the U.S. Environmental Protection Agency, 109 T.W. Alexander Drive, Research Triangle Park, NC 27709, USA

[4]Denver Federal Center, U.S. Geological Survey, Box 25046, MS 973, Denver, CO 80225, USA

[5]National Homeland Security Research Center, Threat and Consequence Assessment Division, U.S. Environmental Protection Agency, Ronald Reagan Building, MC 8801RR, 1200 Pennsylvania Avenue, NW, Washington, DC 20460, USA

ABSTRACT

Soil geochemical data from sample sites in counties that reported occurrences of anthrax in wildlife and livestock since 2000 were evaluated against counties within the same states (MN, MT, ND, NV, OR, SD and TX) that did not report occurrences. These data identified the elements, calcium (Ca), manganese (Mn), phosphorus (P) and strontium (Sr), as having statistically significant differences in concentrations between county type (anthrax occurrence*versus* no occurrence). Tentative threshold values of the lowest concentrations of each of these elements (Ca = 0.43 wt %, Mn = 142 mg/kg, P = 180 mg/kg and Sr = 51 mg/kg) and average concentrations (Ca = 1.3 wt %, Mn = 463 mg/kg, P = 580 mg/kg and Sr = 170 mg/kg) were identified from anthrax-positive counties as prospective investigative tools in determining whether an outbreak had "potential" or was "likely" at any given geographic location in the contiguous United States.

INTRODUCTION

B. anthracis infections in wildlife and livestock have been recognized as a critically important disease in the United States for over 200 years. Historical data on environmental, weather/climate and geographical factors that influence the occurrence of these infections are well known and include; (1) warm seasons during dry periods that follow moderate to heavy precipitation events (weather/climate); (2) regions containing post-flood organic detritus and/or short dry grazing grasses (environmental); and (3) topological lows, such as waterholes or riverbanks, calcareous and alluvial soils with elevated nutrient content and pH values greater than 6.0 (geology). Other geological factors that may influence *B. anthracis* outbreak occurrence, as noted through *in vivo* or *in vitro* observations, are elevated phosphate (which results in higher protective antigen production), magnesium, sodium, copper, zinc (needed for lethal factor production) and manganese (typically found in very low concentrations in calcareous soils and needed for gene regulation of exotoxins and antibiotics) [1,2,3,4,5].

There are over 140 strains of *Bacillus anthracis*, and all pathogenic strains carry both pX01 and pX02 virulence plasmids [6]. Two separate groups of *B. anthracis*, the "Ames" and Western North America (WNA) clades, are responsible for wildlife and livestock anthrax outbreaks in North America. Animal outbreaks of anthrax are a common occurrence in the contiguous United States, and they are typically constrained to a few geographical regions (e.g., Texas, Minnesota, Montana and the Dakotas). The "Ames" or "Ames-like" clade has caused periodic outbreaks in southern Texas and is believed to have been introduced through the importation of infected livestock during European colonization [7,8]. The WNA clade is genetically most similar to isolates of the Eurasian clade and account for ~89% of non-human cases in North America [7]. It is believed that the WNA clade was introduced to the Americas by human migration across the Bering Strait that occurred prior to ~11,000 years ago when the land bridge between Asia and North America last closed at the end of the Younger Dryas [7,9,10]. Genetic analyses of WNA clade isolates show evidence of a north to south distribution pattern that is rooted in northern Canada [7]. Costs associated with outbreaks can be significant. The 2005 North Dakota outbreak was estimated to have cost ~$650 thousand U.S. dollars (costs associated with activities, such as surveillance, diagnosis, immunization and disposal) [11]. Similarly, the periodic large outbreaks that affect bison and other wildlife in Canada are believed to cost

~$500 thousand Canadian dollars per episode, and various Canadian agencies spend an estimated $15 thousand to $26 thousand per year on aerial carcass surveillance [12]. Even small outbreaks can significantly impact the economic well-being of the livestock industry, where profit margins are based on low expected annual herd losses [13].

Given the geographic restriction of most annually-occurring cases and outbreaks of anthrax in the contiguous United States, geochemical data obtained by the U.S. Geological Survey's (USGS) "North American Soil Geochemical Landscapes Project" were evaluated in collaboration with the Environmental Protection Agency (EPA) to determine which elements may influence the background distribution of this pathogen. These data may help decision makers better prepare for and mitigate potential or actual outbreak events and provide an accurate graphical representation of areas within the contiguous United States that favor the natural propagation of this species.

EXPERIMENTAL SECTION

Sample Sites and Geochemical Data

Using a generalized random tessellation stratified design for sample site selection, 4,857 sample sites (~1 site per 1,600 km²) were utilized for the USGS North American Soil Geochemical Landscapes Project, and 209 of those sites were utilized in this study [14]. In a major geochemical mapping project such as this, the quality of chemical analyses is of utmost importance. Reimann *et al.* (2008) recommend the following five quality control (QC) procedures [15]:

   Collection and analysis of field duplicates;

- Randomization of samples prior to analysis;
- Insertion of international reference materials (RMs);
- Insertion of project standards; and
- Insertion of analytical duplicates of project samples.

In this project, field duplicates were not collected. This approach was evaluated during the pilot studies (Smith *et al.*, 2009) and reported on by Garrett (2009) [16,17]. Based on the results of the pilot studies, it was felt that the additional collection of field duplicates during the national-scale study would not add significantly to the QC analysis and, therefore, was not worth the added expense. The remaining four QC procedures were carried out fully.

To estimate trueness as measured in terms of bias, one or more standards consisting of both international RMs and internal project standards were analyzed with the project samples. In this project, trueness estimation was done on three separate levels. The USGS contract laboratory analyzed an RM with every batch of 48 samples. At the second tier, the USGS QC officer inserted at least one RM between every batch of 20–30 samples. The USGS principal investigator for the project (David B. Smith) initiated the final QC tier, which included the insertion of two blind RMs within each batch of 20–30 samples. Precision was assessed both by repeated analyses of RMs and by replicate analyses of real project samples. Quality control samples (RMs and analytical duplicates) constituted approximately 12% of the total number of samples analyzed. A complete discussion of the QC protocols used in this project, including detailed tables of bias and precision, is given in Smith *et al.*(2013) [14].

In short, the <2-mm fraction of each sample that was collected from a depth of 0 to 5 cm below the soil surface was analyzed for aluminum (Al), arsenic (As), calcium (Ca), iron (Fe), mercury (Hg), potassium (K), magnesium (Mg), sodium (Na), sulfur (S), titanium (Ti), silver (Ag), barium (Ba), beryllium (Be), bismuth (Bi), cadmium (Cd), cerium (Ce), cobalt (Co), chromium (Cr), cesium (Cs), copper (Cu), gallium (Ga), indium (In), lanthanum (La), lithium (Li), manganese (Mn), molybdenum (Mo), niobium (Nb), nickel (Ni), phosphorus (P), lead (Pb), rubidium (Rb), antimony (Sb), scandium (Sc), selenium (Se), tin (Sn), strontium (Sr), tellurium (Te), thorium (Th), thallium (Tl), uranium (U), vanadium (V), tungsten (W), yttrium (Y) and zinc (Zn) [14]. Elemental concentrations were reported as weight percent (wt % = Al, Ca, Fe, K, Mg, Na, Ti and S) or milligrams per kilogram (mg/kg) [14].

B. Anthracis Case and Outbreak Data by State County, 2000–2013

Figure 1 illustrates state counties reporting outbreaks or cases of anthrax in agricultural animals/wildlife since 2000 (red counties). States utilized for statistical analyses included Minnesota, Montana, North Dakota, Nevada, Oregon, Texas and South Dakota. State county outbreak and case data were compiled from state animal health organizations and the National Animal Health Reporting System [18]. Geochemical sample sites (USGS Geochemical Landscape Project sample site numbers presented in data tables [14]) were chosen within each county (Table 1). The following anthrax-positive counties were utilized for statistical evaluation: (1) Minnesota: Clay, Kittson, Lake of the Woods, Marshall Pennington, Polk and Roseau; (2) Montana: Gallatin, Sheridan and Roosevelt; (3) Nevada: Washoe; (4) North Dakota: Barnes, Cass, Grand Forks, Nelson, Pembina, Stark, Steele and Traill; (5) Oregon: Klamath; (6) South Dakota: Aurora, Brown, Brule, Buffalo, Charles Mix, Corson, Day, Dewey, Hand, Hughes, Hyde, Lyman, Marshall, Mellette, Potter, Spink, Tripp and Walworth; and (7) Texas: Edwards, Irion, Kinney, McCulloch, Real, Sutton, Uvalde and Val Verde. In summary, there were 120 sample sites located within these 46 counties.

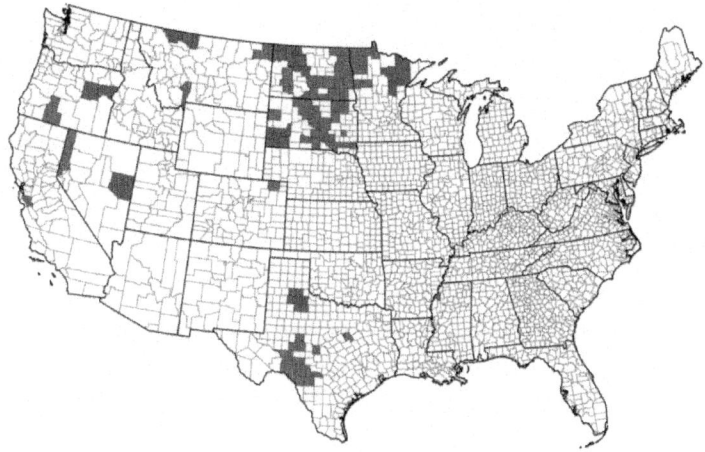

Figure 1: Counties (red) in the contiguous United States reporting cases and/or outbreaks of agricultural/wildlife anthrax since 2000. Counties with no reported cases (blue) where sample sites were utilized for geochemical statistical analyses *versus* those sample sites in counties in the same state that reported cases and/or outbreaks

Table 1: State, county and U.S. Geological Survey's (USGS) sample site data [14]

State	Counties (total number)	USGS Geochemical Landscape Project sample site numbers (numbers grouped by county (total number)
	**State Counties Reporting Outbreaks or Cases of Anthrax in Livestock or Wildlife, 2000–2013 Utilized for Statistical Evaluation**	
Minnesota	Clay, Kittson, Lake of the Woods, Marshall, Pennington, Polk and Roseau (7)	2265, 6361, 857, 9094, 7129, 10969, 4825, 4953, 8921, 9177, 12121, 729, 1753, 3545, 6617, 7641 and 2009 (17)
Montana	Gallatin, Sheridan and Roosevelt (3)	2798, 3310, 5246, 6974, 11070, 1854, 7742 and 12414 (8)
Nevada	Washoe (1)	671, 1503, 2719, 3551, 5791, 6815, 8863, 9695, 10719, 11743 and 12447 (11)
North Dakota	Barnes, Cass, Grand Forks, Nelson, Pembina, Stark, Steele and Traill (8)	601, 1177, 4697, 8793, 3417, 7513, 11609, 5273, 8025, 8345, 9369, 1881, 3966, 5310, 8062, 12441, 2201 and 6297 (18)
Oregon	Klamath (1)	14, 927, 1951, 5023, 6047, 7162, 8186, 10143, 11258, 12282 and 13215 (11)
South Dakota	Aurora, Brown, Brule, Buffalo, Charles Mix, Corson, Day Dewey, Hand, Hughes, Hyde, Lyman, Marshall, Mellette, Potter, Spink, Tripp and Walworth (18)	1224, 8648, 10120, 4185, 8985, 3528, 4808, 13000, 4296, 12744, 1662, 9086, 10878, 12926, 3865, 7961, 3275, 4734, 8830, 10443, 7624, 11720, 1736, 5832, 9928, 6347, 8904, 1113, 2137, 12377, 3723, 638, 6937, 11033, 89, 200, 712, 456, 2504, 10891, 11464, 2841 and 10009 (43)
Texas	Edwards, Irion, Kinney, McCulloch, Real, Sutton, Uvalde and Val Verde (8)	4804, 8900, 12095, 7364, 9656, 3356, 6092, 708, 6596, 11716, 7452 and 12996 (12)
	**State Counties Not Reporting Outbreaks or Cases of Anthrax in Livestock or Wildlife, 2000–2013 Utilized for Statistical Evaluation**	
Minnesota	Aitkin, Itasca and St. Louis (3)	1077, 11317, 12341, 473, 1653, 1689, 3289, 2677, 3701, 4725, 6005, 7797 and 8245 (13)
Montana	Glacier, Toole and Liberty (3)	3502, 11694, 2222, 3246, 7342, 10414, 11438, 6318 and 8366 (9)
Nevada	White Pine (1)	271, 1359, 2015, 2063, 3087, 3407, 4367, 8463, 9231, 9551, 10255, 11279, 11375 and 11999 (14)
North Dakota	Burke, Divide, Mclean, Mountrail, Renville, Ward and Williams (7)	7150, 11246, 12270, 3054, 6462, 7486, 10558, 318, 11326, 2030, 62, 5438, 9534, 3134, 7230 and 13118 (16)
Oregon	Baker and Grant (2)	3342, 6926, 8974, 9742, 10510, 13070, 1102, 1550, 2574, 4174, 4622, 5198, 5646, 9294 and 12366 (15)
South Dakota	Custer, Fall River, Pennington and Shannon (4)	1675, 4811, 11147, 3979, 13003, 651, 4123, 4747, 5771, 8395 and 12491 (11)
Texas	Briscoe, Cottle, Dickens, Floyd, Hall, King and Motley (7)	4735, 3967, 7039, 11135, 191, 11391, 8063, 11647, 6015, 10111 and 4287 (11)

The anthrax-negative counties utilized for statistical evaluation (these were chosen randomly without knowledge of site geochemistry from each relevant state after the anthrax positive counties were mapped) included: (1) Minnesota: Aitkin, Itasca and St. Louis; (2) Montana: Glacier, Toole and Liberty; (3) Nevada: White Pine; (4) North Dakota: Burke, Divide, Mclean, Mountrail, Renville, Ward and Williams; (5) Oregon: Baker and Grant; (6) South Dakota: Custer, Fall River, Pennington and Shannon; and (7) Texas: Briscoe, Cottle, Dickens, Floyd, Hall, King and Motley. In summary, there were 89 sample sites located within these 27 counties.

Statistics

The non-parametric Mann–Whitney U test was utilized to evaluate differences in geochemistry between counties where anthrax outbreaks or cases had been reported since the year 2000 and counties within the same states where no cases were noted for the same time period using SPSS (IBM, Tampa, FL, USA) [19]. In the USGS Geochemical Landscape Project element concentration data set, there are values expressed as below minimum detection limits (MDL) for certain elements (Ag = 189 of 209, Cs = 170/209, Cd = 19/209, S = 2/209, Se 67/209 and Te = 198/209 data points). For statistical analyses, those values were set at the MDL for the respective elements (e.g., <1 is set at 1).

RESULTS AND DISCUSSION

Comparing 120 sample sites from 46 counties (seven states, MN, MT, NV, ND, OR, TX and SD) that had reported anthrax outbreaks or cases to 89 sites from 27 counties (same states) that did not report outbreaks or cases resulted in the identification of seven elements with statistically significant differences in their respective concentrations (Table 2, all counties, Column 2). These elements included Ca ($p = 0.006$), Nb ($p = 0.035$), Ni ($p = 0.028$), P ($p = 0.028$), S ($p = 0.002$), Sn ($p = 0.024$) and Sr ($p = 0.041$). With the exception of Nb and Sr, the total state average of elemental concentrations was higher in anthrax-positive counties. When the elements were looked at individually, several trends emerged.

Strontium

When contrasting the elements by each state, only Sr had average concentrations that were higher in all anthrax-positive counties *versus* anthrax-negative counties, and the lowest observed concentration was 116 mg/kg. Strontium data were significantly different in three of the seven states.

Calcium

These concentrations were similar in both types of counties, with only one instance where average concentrations in negative counties exceeded positive counties, and that was in NV at 5.05 and 3.03 wt %, respectively. This anomaly can be explained in that the average concentrations in both the negative and positive counties were the second and third overall highest average concentrations in comparison to the data obtained from each of the other evaluated states. Overall, calcium data were significantly different between county types in three of the seven states.

Table 2: The significance (Mann–Whitney U test p-values; in bold where <0.05) of elemental concentrations (averages in brackets [#/#] where there was an overall or greater than two state significant p-values) in counties reporting outbreaks or cases of anthrax in livestock and wildlife *versus* counties that have not reported outbreaks or cases of anthrax since the year 2000

Element	All Counties 46 (120)/27 (89)	Texas 8 (12)/7 (11)	N. Dakota 8 (18)/7 (16)	S. Dakota 18 (43)/4 (11)	Minnesota 7 (17)/3 (13)	Nevada 1 (11)/1 (14)	Oregon 1 (11)/2 (15)	Montana 3 (8)/3 (9)
Al	0.791					0.000	0.001	0.034 *
	[5.9/5.3]	[4.2/3.9]	[4.6/4.7] *	[5.1/4.8]	[4.5/4.1]	[8.4/5.6]	[9.6/8.0]	[4.7/5.7]
As		0.018						
Ba	0.223 *	0.001 *	0.001 *		0.021 *	0.001		
	[582/634]	[283/444]	[556/627]	[691/755] *	[485/520]	[928/672]	[528/682] *	[599/740] *
Be								0.041 *
Bi		0.002				0.015 *		
Ca	0.006	0.000		0.004	0.005			
	[3.3/1.8]	[10.4/1.1]	[1.8/1.4]	[1.3/0.7]	[2.2/1.0]	[3.0/5.1] *	[2.6/2.5]	[1.9/0.9]
Co						0.000	0.023 *	
Cr		0.019						
Cu								0.034 *
Fe						0.000		
Ga						0.000		
K		0.019*						0.012 *
La						0.004 *	0.023 *	
Li						0.001 *		
Ln		0.019						
Mn	0.072	0.006	0.045	0.01	0.004 *	0.014		
	[761/702]	[530/304]	[783/602]	[1024/530]	[463/1144]	[925/569]	[1120/1343]	[487/424]
Mo	0.128 *	0.001		0.009 *	0.002 *			
	[0.9/1.0]	[1.1/0.6]	[0.8/0.8]	[1.2/1.5]	[0.4/0.8]	[1.3/1.2]	[0.9/1.3] *	[0.7/1.1] *
Na	0.693	0.001 *		0.001			0.02	
	[1.2/0.9]	[0.2/0.5]	[0.8/0.9]	[0.9/0.6]	[1.1/1.1] *	[2.2/1.0]	[2.3/1.6]	[0.9/0.9]
Nb	0.035 *	0.021					0.000 *	
	[7.9/9.0]	[10.1/7.5]	[6.8/6.8]	[8.8/9.9]	[5.7/6.1] *	[10.2/12.7] *	[6.2/10.8]	[7.7/9.3] *

Element	All Counties 46 (120)/27 (89)	Texas 8 (12)/7 (11)	N. Dakota 8 (18)/7 (16)	S. Dakota 18 (43)/4 (11)	Minnesota 7 (17)/3 (13)	Nevada 1 (11)/1 (14)	Oregon 1 (11)/2 (15)	Montana 3 (8)/3 (9)
Ni	0.028			0.01				
	[21/18]	[15/12]	[19/17]	[25/15]	[15/14]	[16/14]	[42/30]	[14/20] *
P	0.028	0.003	0.002	0.01				
	[761/692]	[580/330]	[652/518]	[737/566]	[675/620]	[818/886] *	[1,203/1,099]	[658/827] *
Pb	0.190 *	0.004		0.000 *	0.013 *	0.023 *	0.043 *	
	[16/18]	[16/13]	[14/13]	[16/21]	[14/18]	[16/21]	[11/13]	[14/25] *
Rb	0.21 *					0.001 *	0.012 *	0.034 *
	[58/70]	[69/64]	[64/64]	[69/67]	[58/60] *	[61/106]	[27/43]	[61/84]
S	0.002 *	0.000						
	[0.05/0.06]	[0.06/0.02]	[0.05/0.04]	[0.05/0.04]	[0.04/0.04]	[0.06/0.03]	[0.03/0.03]	[0.06/0.19] *
Sb		0.000					0.045 *	
Sn	0.024			0.003 *		0.029 *		
	[1.53/1.34]	[3.67/1.19]	[1.03/0.93]	[1.18/1.40]	[0.89/0.95]	[1.46/1.81]	[1.39/1.75]	[1.09/1.34]
Sr	0.041			0.003		0.000	0.000	
	[250/169]	[116/86]	[165/154]	[170/137]	[189/179]	[457/262]	[495/229]	[157/140]
Th						0.001 *	0.012 *	
Ti						0.000		
Tl	0.498 *	0.029				0.002 *	0.013 *	0.036 *
	[0.4/0.47]	[0.46/0.35]	[0.44/0.43]	[0.52/0.55] *	[0.35/0.36] *	[0.37/0.61]	[0.23/0.35]	[0.44/0.62]
U	0.837 *	0.006	0.007 *					0.023 *
	[1.90/1.91]	[1.70/1.71] *	[1.92/1.45]	[1.99/2.40]	[2.11/1.14]	[2.05/2.42] *	[1.42/1.69] *	[1.80/2.58]
V						0.000		0.014 *
W		0.04		0.015 *				
Y		0.018					0.008 *	
Zn								0.021 *

Notes: Numbers under column titles = the number of counties with anthrax cases (the total number of sample sites in those counties used for analyses)/the number of counties with no cases (the total number sample sites in those counties used for analyses). [#/#], [the average concentration in counties with reported cases/average concentration in counties with no reported cases]. * = lower concentration in anthrax-positive counties. Elemental concentrations are reported as weight percent (wt % = Al, Ca, Fe, K, Mg, Na, Ti and S) or mg/kg [14]. Elements Cd, Ce, Hg, Mg and Sc did not show significance in any of the states and were not included to simplify the table.

Phosphorus

Phosphorus concentration averages in NV (886 mg/kg) and MT (827 mg/kg) were greatest in negative counties, but these concentrations were the third and fourth highest overall concentrations in comparison to the data obtained from the other states. Overall, P data were significantly different in three of the seven states.

Nickel

Average Ni concentrations by state, with the exception of MT, were higher in anthrax-positive counties. The Ni concentrations in the MT counties averaged 20 mg/kg, which was the fourth highest overall. The only significant difference in Ni concentrations by state occurred in SD.

Niobium

Significant differences in total Nb concentrations occurred with only two states showing contrasting data, TX and OR, with average concentrations higher in anthrax-positive counties and in anthrax-negative counties, respectively.

Manganese

Manganese concentrations, while not significant for the total data set ($p = 0.07$), were significant when contrasting counties in TX, ND, SD, MN and

NV. Only in MN was a significant difference noted where the Mn average concentration was greater in negative counties, and in this instance, the negative county average was the second highest observed (1144 mg/kg) across all states. Elevated concentrations such as this may mask a relationship.

Sulfur

The total S significant difference (high concentrations in negative counties) occurred over a small concentration range (0.02 to 0.19 wt %), and the only state-level significant difference that occurred was with the TX data set, which was opposite (high concentrations in positive counties) of the total.

Other Elements

Similar to the observation with sulfur, the total Sn significant difference (high concentrations in positive counties) was opposite that observed with the two state-level data sets. Several other elements, such as Al, Ba, Mo, Na, Pb, Rb and Tl, exhibited significant differences in multiple or individual states, but in many cases, one state produced a significant difference in anthrax-positive counties and, in another, in anthrax-negative counties. Cesium data produced a significant p-value below 0.05, but this was dismissed, due to the fact that 170 of the 209 data points were below the MDL. Of the remaining four elements (Ag, Cd, Te and Se) with MDL data, none produced p-values below 0.05.

Figure 1 illustrates the counties used for statistical analyses and the data (Mann–Whitney U p-values and, in relevant cases, the average elemental concentrations) are listed in Table 2. Of the 40 elements screened, seven (Ca, Nb, Ni, P, S, Sn and Sr) gave significant differences when samples from all seven states were evaluated as a whole. Of these, eight were positive (meaning the concentration was higher in anthrax counties) significant differences and one (Nb) was negative (the concentration was lower in anthrax counties). The Nb differences resulted in both negative (OR) and positive (TX) "by state" results, questioning the strength and/or validity of this "total" observation. The overall differences in concentrations of other elements, such as Ni and S, also resulted in both significant negative and positive results, and thus, the overall observation is either weak or not valid. The significant difference with Ni is also considered weak given that this was derived from a single positive difference (p = 0.01) that was observed within the SD sample set. This observation was also noted with the S data. It may be that one of these or other elements do contribute to virulence, but further research is needed to determine the potential role and threshold concentrations. The remaining three overall positive differences (Ca, P and Sr) had significant p-values in at least three of the seven states for each element. For Mn, there was one negative (due to the second highest average concentration at 1144 mg/kg, relative to the overall seven-state data set average of 808 mg/kg) and four significant positive state data. Manganese was selected for inclusion in the group of selected relevant elements (Ca, Mn, P and Sr) given the predominance of significantly positive state data and the skew produced by the lone negative. The regional distribution and concentration ranges for these four elements (Ca, Mn, P and Sr) and Zn (an element required for the lethal factor) are illustrated in Figure 2. Calcium, Mn and P have also been recognized as elements influencing the growth and/or virulence of this pathogen [1,5,20,21]. Other elements that have been reported to influence this pathogen include Na and S [2,4], and both of these elements resulted in at least one significantly positive state data set (Table 2). Also of note are elements, such as Ba and Rb (both close neighbors to calcium and strontium in the periodic table), which produced multi-state negative significance data, that may inhibit virulence by mechanisms, such as mimicking a critical virulence element [22]. In this case, the probability of conversion is suppressed in geographic regions where the mimicking element exceeds a given threshold concentration. It is interesting (as can be observed in Figure 3) that the concentrations of both of these elements are relatively low in many of the anthrax-positive counties of ND, SD, MN and TX.

Using concentrations observed at sample sites in the states listed in Table 2 for Ca, Mn, P and Sr, several tentative threshold concentrations can be selected for each element in regard to the likelihood of an outbreak occurring at a given location. As an example, the minimum concentration observed in any of these state counties for Ca is 0.43 wt %, and the lowest significant average listed in Table 2 is 1.3 wt %. These concentrations can be utilized as putative thresholds for an investigative tool to determine the likelihood of a naturally occurring outbreak being "potential" at 0.43 wt % or above and "likely" at 1.3 wt % or above. Similarly, "potential" and "likely" thresholds can also be set for Mn (144 and 463 mg/kg), P (180 and 580 mg/kg) and Sr (51 and 170 mg/kg). Figure 4 illustrates those sample sites where those upper or "likely" concentration levels occurred both individually and in combination.

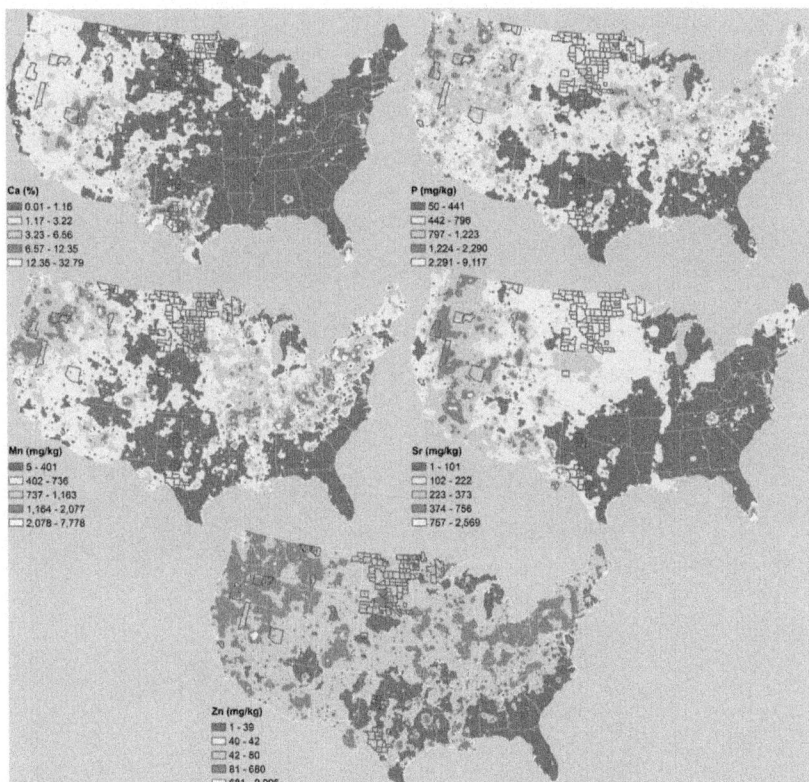

Figure 2: Calcium, phosphorus, manganese, strontium and zinc soil concentration gradient maps for the contiguous United States. Red counties = cases and/or outbreaks of agricultural/wildlife anthrax since 2000. Blue counties = no reported cases and utilized for geochemical statistical comparisons with red counties.

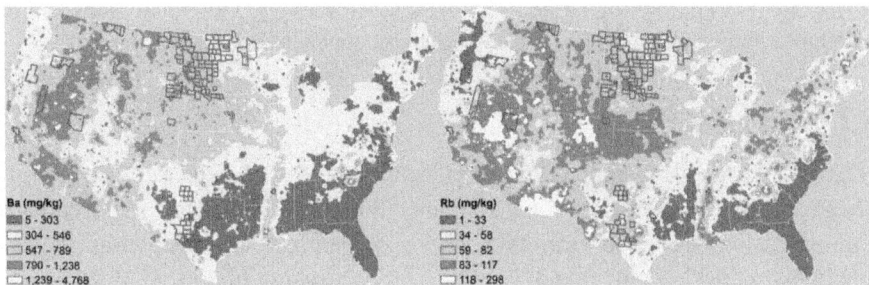

Figure 3: Barium and rubidium soil concentration gradient maps for the contiguous United States. Red counties = cases and/or outbreaks of agricultural/wildlife anthrax since 2000. Blue counties = no reported cases and utilized for geochemical statistical comparisons with red counties.

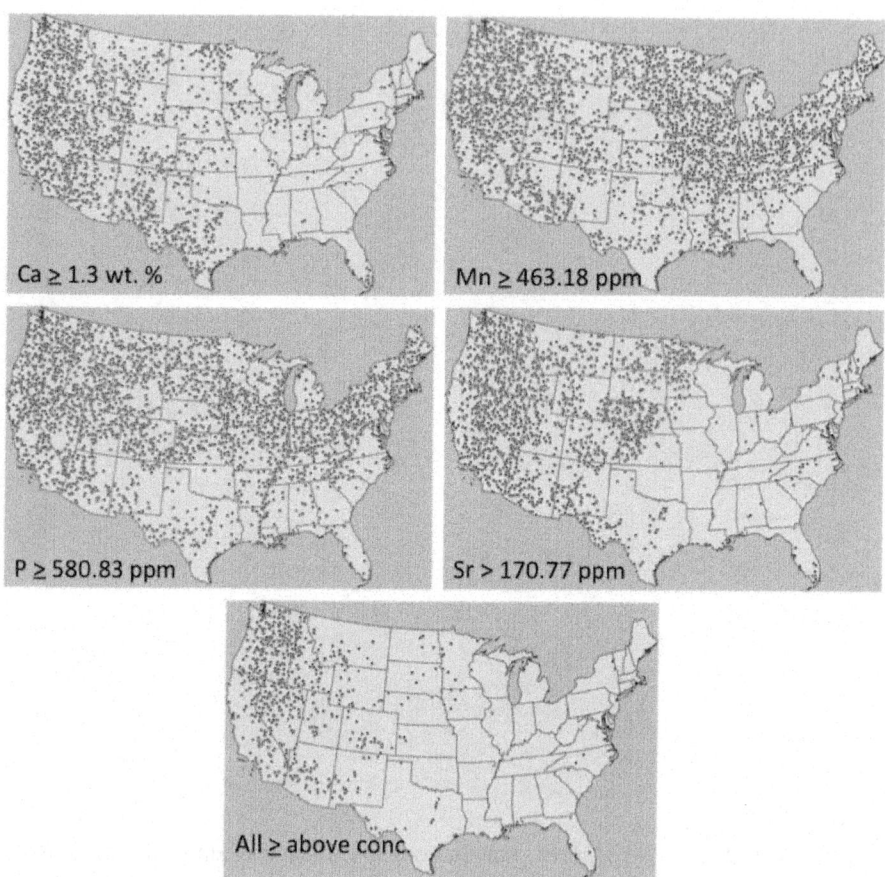

Figure 4: USGS Geochemical Landscape Project sample sites where the average statistically significant concentrations of Ca, Mn, P and Sr were equal to or exceeded 1.3 wt %, 463 mg/kg, 580 mg/kg and 170 mg/kg, respectively. Individual maps and one combined showing the sites where each of these concentrations occurred.

CONCLUSIONS

The evaluation of geochemical data from a series of selected sample sites in seven states identified four elements that had significant differences in concentrations between anthrax-positive and anthrax-negative counties. The elements were Ca, Mn, P and Sr, which in part match historical observations. Tentative threshold values based on the lowest concentrations and the lowest average concentrations of each of these elements, in the anthrax positive-counties utilized in this study, were identified for use as prospective tools for determining whether or not a naturally occurring outbreak had "potential" or was "likely" at any given geographic location. While these elemental threshold values are preliminary in nature, they present an investigative tool that can be refined through future high-resolution studies that need to be conducted in and around "endemic" areas. The USGS data set is a valuable tool that can be used to determine the background distribution of pathogens in soils of the contiguous United States. Being able to predict the natural occurrence of this agent may help guide animal and public health planning and response efforts. These data also provide insight to assist in environmental remediation decisions following a suspected outbreak or release of this agent and, overall, provide a roadmap forward for investigating the natural background occurrence of other select agents.

ACKNOWLEDGMENTS

This project was a joint USGS/USEPA (through its Office of Research and Development) collaboration under EPA IA# DW14957748. The authors would like to thank Sarah Perkins formerly of the USEPA for her help and assistance on this project. This content has been peer and administratively reviewed and has been approved for publication as a joint USGS and USEPA publication. Note that approval does not signify that the contents necessarily reflect the views of the USEPA or the USGS, but rather the authors. The use of trade names is for descriptive purposes only and does not imply endorsement by the U.S. Government.

REFERENCES

1.   Weinberg, E.D. The Influence of soil on infectious-disease. Experientia 1987, 43, 81–87.

2.   Griffin, D.W.; Petrosky, T.; Morman, S.A.; Luna, V.A. A survey of the occurrence of Bacillus anthracis in North American soils over two long-range transects and within post-Katrina New Orleans. Appl. Geochem. 2009, 24, 1464–1471.

3.   Kochi, S.K.; Schiavo, G.; Mock, M.; Montecucco, C. Zinc content of the Bacillus anthracis lethal factor. Fems Microbiol. Lett. 1994, 124, 343–348.

4.   Hugh-Jones, M.; Blackburn, J. The ecology of Bacillus anthracis. Mol. Aspects Med. 2009, 30, 356–367.

5.   Wright, G.G.; Angelety, L.H.; Swanson, B. Studies on immunity in anthax. XII. Requirement for phosphate for elaboration of protective antigen and its partial replacement by charcoal. Infect. Immun. 1970, 2, 772–777.

6. Qi, Y.; Patra, G.; Liang, X.; Williams, L.E.; Rose, S.; Redkar, R.J.; DelVecchio, V.G. Utilization of the rpoB gene as a specific chromosomal marker for real-time PCR detection of Bacillus anthracis. Appl. Environ. Microb. 2001, 67, 3720–3727.

7. Kenefic, L.J.; Pearson, T.; Okinaka, R.T.; Schupp, J.M.; Wagner, D.M.; Ravel, J.; Hoffmaster, A.R.; Trim, C.P.; Chung, W.K.; Beaudry, J.A.; Foster, J.T.; Mead, J.I.; Keim, P. Pre-Columbian origins for North American anthrax. PLoS One 2009, 4, e4813.

8. Van Ert, M.N.; Easterday, W.R.; Huynh, L.Y.; Okinaka, R.T.; Hugh-Jones, M.E.; Ravel, J.; Zanecki, S.R.; Pearson, T.; Simonson, T.S.; U'Ren, J.M.; et al. Global Genetic Population Structure of Bacillus anthracis. PLoS One 2007, 2, e461.

9. Elias, S.A.; Short, S.K.; Nelson, C.H.; Birks, H.H. Life and times of the Bering land bridge.Nature 1996, 382, 60–63.

10. Williams, R.C.; Steinberg, A.G.; Gershowitz, H.; Bennett, P.H.; Knowler, W.C.; Pettitt, D.J.; Butler, W.; Baird, R.; Dowdarea, L.; Burch, T.A.; et al. Gm Allotypes in Native Americans—Evidence for 3 Distinct Migrations across the Bering Land-Bridge. Am. J. Phys. Anthropol. 1985,66, 1–19.

11. Mongoh, M.N. Characterization of Anthrax Occurrence in North Dakota: Determinants, Management Strategies, and Economic Impacts. Ph.D. Thesis, North Dakota State University of Agriculture and Applied Sciences, Fargo, North Dakota, USA, September 2007.

12. Salb, A.; Stephen, C.; Ribble, C.; Elkin, B. Descriptive epidemiology of detected anthrax outbreaks in wild wood bison (bison bison athabascae) in northern Canada,1962–2008. J. Wildl. Dis. 2014, 50, 459–468.

13. Ramsey, R.; Doye, D.; Ward, C.; McGrann, J.; Falconer, L.; Bevers, S. Factors affecting beef cow-herd costs, production, and profits. J. Agr. Appl. Econ. 2005, 37, 91–99.

14. Smith, D.B.; Cannon, W.F.; Woodruff, L.G.; Solano, F.; Kilburn, J.E.; Fey, D.L. Geochemical and Mineralogical Data for Soils of the Conterminous United States; U.S. Geological Survey: Reston, VA, USA, 2013. Available online: http://pubs.usgs.gov/ds/801/ (accessed on 12 May 2014).

15. Reimann, C.; Filzmoser, P.; Garrett, R.G.; Dutter, R. Statistical Data Analysis Explained; John Wiley & Sons: Chichester, UK, 2008.

16. Smith, D.B.; Woodruff, L.G.; O'Leary, R.M.; Cannon, W.F.; Garrett, R.G.; Kilburn, J.E.; Goldhaber, M.B. Pilot studies for the North American Soil Geochemical Landscapes Project—Site selection, sampling protocols,

analytical methods, and quality control protocols. Appl. Geochem.2009, 24, 1357–1368.

17. Garrett, R.G. Relative spatial soil geochemical variability along two transects across the United States and Canada. Appl. Geochem. 2009, 24, 1405–1415.

18. Animal and Plant Health Inspection Service, United States Department of Agriculture. Available online: http://www.aphis.usda.gov/wps/portal/aphis/home/ (accessed on 8 August 2014).

19. Dytham, C. Choosing and Using Statistics, A Biologist's Guide; Blackwell Science: Oxford, UK, 1999.

20. Pasteur, L. On the etiology of anthrax. Comptes Rendus Seances Acad. Sci. 1880, 91, 86–94.

21. Van Ness, G.; Stein, C.D. Soils of the United States favorable for anthrax. J. Am. Vet. Med. Assoc. 1956, 128, 7–12.

22. Heldman, E.; Levine, M.; Raveh, L.; Pollard, H.B. Barium ions enter chromaffin cells via voltage-dependent calcium. J. Biol. Chem. 1989, 264, 7914–7920.

Chapter 5

RECENT ADVANCES AND FUTURE CHALLENGES FOR ARTIFICIAL NEURAL SYSTEMS IN GEOTECHNICAL ENGINEERING APPLICATIONS

Mohamed A. Shahin,[1] Mark B. Jaksa,[2] and Holger R. Maier[2]

[1]Department of Civil Engineering, Curtin University of Technology, Perth, WA 6845, Australia

[2]School of Civil, Environmental and Mining Engineering, University of Adelaide, Adelaide, SA 5005, Australia

ABSTRACT

Artificial neural networks (ANNs) are a form of artificial intelligence that has proved to provide a high level of competency in solving many complex engineering problems that are beyond the computational capability of classical mathematics and traditional procedures. In particular, ANNs have been applied successfully to almost all aspects of geotechnical engineering problems. Despite the increasing number and diversity of ANN applications in geotechnical engineering, the contents of reported applications indicate that the progress in ANN development and procedures is marginal and not moving forward since the mid-1990s. This paper presents a brief overview of ANN applications in geotechnical engineering, briefly provides an overview of the operation of ANN modeling, investigates the current research directions of ANNs in geotechnical engineering, and discusses some ANN modeling issues that need further attention in the future, including model robustness; transparency and knowledge extraction; extrapolation; uncertainty.

INTRODUCTION

Artificial neural networks (ANNs) are well suited to model the complex behavior of most geotechnical engineering materials which, by their very nature, exhibit extreme variability. ANNs have also demonstrated superior predictive ability when compared with traditional methods. Since the early

1990s, ANNs have been applied successfully to virtually every problem in geotechnical engineering. In this section, post-2001 applications of ANNs in geotechnical engineering are briefly examined, and interested readers are referred to Shahin et al. [1], where the pre-2001 papers are reviewed in some detail.

The behavior of deep (pile) and shallow foundations in soils is complex, uncertain, and not yet entirely understood. This fact has encouraged many researchers to apply the ANN technique to the prediction of the behavior of foundations. For example, ANNs have been used extensively for modeling the axial and lateral load capacities of deep foundations in compression and uplift, including driven piles [2–6], drilled shafts [7,8], and ground anchor piles [9, 10]. The prediction of behavior of shallow foundations has also been investigated, including settlement estimation [11–16] and bearing capacity [17–19].

Classical constitutive modeling based on elasticity and plasticity theories has limited capability to simulate properly the behavior of geomaterials. This is attributed to reasons associated with the formulation complexity, idealization of material behavior, and excessive empirical parameters [20]. In this regard, many neural networks have been proposed as a reliable and practical alternative to model the constitutive monotonic and hysteretic behavior of geomaterials [21–29].

Geotechnical properties and behavior of soils are controlled by factors such as mineralogy; fabric; pore water, and the interactions of these factors are difficult to establish solely by traditional statistical methods due to their interdependence [30]. Based on the application of ANNs, methodologies have been developed for estimating several soil properties, including the preconsolidation pressure [31], shear strength and stress history [30, 32–37], swell pressure [38, 39], lateral earth pressure [40], compaction characteristics and permeability [41, 42], soil composition and classification [43, 44], and properties of soil dynamics [45, 46]. Liquefaction during earthquakes is one of the very dangerous ground failure phenomena that can cause a large amount of damage to most civil engineering structures. Although the liquefaction mechanism is well known, the prediction of liquefaction potential is very complex [47]. This fact has attracted many researchers to investigate the applicability of ANNs for predicting liquefaction [47–55].

Other applications of ANNs in geotechnical engineering include earth retaining structures [56], dams [57,58], blasting [59], mining [60], environmental geotechnics [61], rock mechanics [62–67], site characterization [68], tunnels and underground openings [69–74], slope stability and landslides [71, 75–79], and deep excavation [80].

BRIEF OVERVIEW OF ARTIFICIAL NEURAL NETWORKS

Many authors have described the structure and operation of ANNs (e.g., [81, 82]), and whilst a comprehensive description of ANNs is beyond the scope of this paper, it is useful to provide a brief overview. ANNs are a data driven artificial intelligence approach that attempts to mimic, in a very simplistic way, the cognition capability of the human brain. ANNs learn by examples of data inputs and outputs presented to them so that the subtle functional relationships among the data are captured, even if the underlying relationships are unknown or the physical meaning is difficult to explain. This is in contrast to most traditional empirical and statistical methods, which need prior knowledge about the nature of the relationships among the data. This is one of the main benefits of ANNs when compared with most empirical and statistical methods.

Typically, the architecture of ANNs consists of a series of processing elements (PEs), or nodes, that are usually arranged in layers: an input layer, an output layer, and one or more hidden layers, as shown in Figure 1.

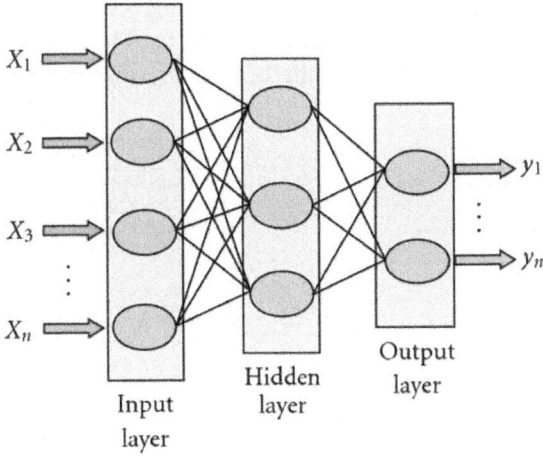

Artificial neural network

(a)

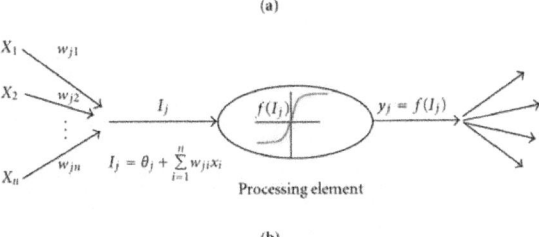

(b)

Figure 1: Typical structure and operation of ANNs.

The input from each PE in the previous layer x_i is multiplied by an adjustable connection weight w_{ji}. At each PE, the weighted input signals are summed and a threshold value θ_j is added. This combined input Ij is then passed through a nonlinear transfer function $f(\cdot)$ to produce the output of the PE y_j. The output of one PE provides the input to the PEs in the next layer. This process is summarized in (1) and (2) and illustrated in Figure 1.1.

$$I_j = \sum w_{ji}x_i + \theta_j \quad \text{summation,} \tag{1}$$

$$y_j = f(I_j) \quad \text{transfer.} \tag{2}$$

The propagation of information in an ANN starts at the input layer, where the input data are presented. The network adjusts its weights on the presentation of a training data set and uses a learning rule to find a set of weights that will produce the input/output mapping that has the smallest possible error. This process is called "learning" or "training." Once the training phase of the model has been successfully accomplished, the performance of the trained model needs to be validated using an independent testing set. The main steps involved in the development of an ANN, as suggested by Maier and Dandy [83], are illustrated in Figure 2. Several of these steps are discussed in some depth in the following section.

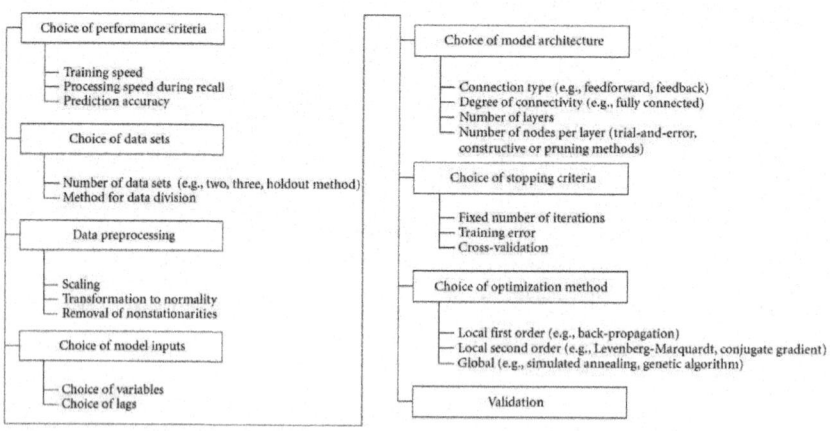

Figure 2: The main steps in ANN model development [83].

CURRENT DEVELOPMENT AND FUTURE DIRECTIONS IN UTILIZATION OF ANNS

One issue that needs to be addressed in order to improve the performance of ANN models is the utilization of a systematic approach in their development.

Such an approach needs to address major factors, including the determination of adequate model inputs, data division and preprocessing, choice of suitable network architecture, careful selection of some internal parameters that control the optimization method, stopping criteria, and model validation. For example, in relation to the second step of choice of data sets, method for data division, Shahin et al. [84] provided guidance using a geotechnical engineering example, and recommended the use of three, statistically consistent but independent data sets, one for each of training, testing, and validation. In this context, Shahin et al. [84] have introduced three approaches so that data division can be carried out in a systematic manner, including trial-and-error, self-organizing maps, and fuzzy clustering. For a detailed treatment of each of the steps in the model development process, interested readers are referred to Shahin et al. [85].

Other key issues in relation to ANN modeling that have received recent attention and require further research in the future include developing approaches that (i) ensure the development of robust models, (ii) increase model transparency and enable knowledge to be extracted from trained ANNs, (iii) improve extrapolation ability, and (iv) deal with uncertainty. Each of these is discussed in what follows.

Model Robustness

Model robustness is the predictive ability of ANN models to generalize over a range of data similar to that used for model training. Kingston et al. [86] stated that if "ANNs are to become more widely accepted and reach their full potential..., they should not only provide a good fit to the calibration and validation data, but the predictions should also be plausible in terms of the relationship modeled and robust under a wide range of conditions." and that "while ANNs validated against error alone may produce accurate predictions for situations similar to those contained in the training data, they may not be robust under different conditions unless the relationship by which the data were generated has been adequately estimated." This is in agreement with the investigation into the robustness of ANNs carried out by Shahin et al. [87] for a case study of predicting the settlement of shallow foundations on granular soils. Shahin et al. [87] found that good performance of ANN models on the data used for model calibration and validation does not guarantee that the models will perform well in a robust fashion over a range of data similar to those used in the model calibration phase. For this reason, Shahin et al. [87] proposed a method to test the robustness of the predictive ability of ANN models by carrying out a sensitivity analysis to investigate the response of ANN model outputs to changes in its inputs. The robustness of the model can then be determined by examining how well model predictions are in agreement with the known underlying physical processes of the problem in hand over a range of inputs. In addition, Shahin et al. [87] advised that the connection weights should be examined as part of the interpretation of ANN model behavior, using, for example, the method suggested by Garson [88]. On the other hand, Kingston et al. [86] adopted the connection weight approach of Olden et al. [89] for a case study in hydrological modeling in order to assess the relationship modeled by the ANNs, as Olden et al. [89] found that this approach provided the best overall methodology for quantifying ANN input importance in comparison to other commonly used methods, though with a few limitations.

Support vector machines (SVMs) are an alternative data-driven modeling approach that is claimed to provide better generalization capabilities and higher accuracy than ANNs and are therefore worth further consideration in relation to achieving improved model robustness [90]. Interested readers are referred to A. T. C. Goh and S. H. Goh [91] for a good overview of this technique. Recent applications of SVMs in the field of geotechnical engineering include the prediction of liquefaction potential [90, 91], analysis of slope stability [92], and modeling friction capacity of driven piles [93].

Model Transparency and Knowledge Extraction

Model transparency and knowledge extraction are the feasibility of interpreting ANN models in a way that provides insights into how model inputs affect outputs. Figure 3 shows the classification of modeling techniques based on colors [94] in which the higher the physical knowledge used during model development, the better the physical interpretation of the phenomenon that the model provides to the user. It can be seen that the color coding of mathematical modeling can be classified into: white-, black-, and grey-box models, each of which can be explained as follows [95]. White-box models are systems that are based on first principles (e.g., physical laws) where model variables and parameters are known and have physical meaning by which the underlying physical relationships of the system can be explained. Black-box models are data-driven or regressive systems in which the functional form of relationships between model variables is unknown and needs to be estimated. Black-box models rely on data to map the relationships between model inputs and corresponding outputs rather than to find a feasible structure of the model input-output relationships. Grey-box models are conceptual systems in which the mathematical structure of the model can be derived, allowing further information of the system behavior to be resolved.

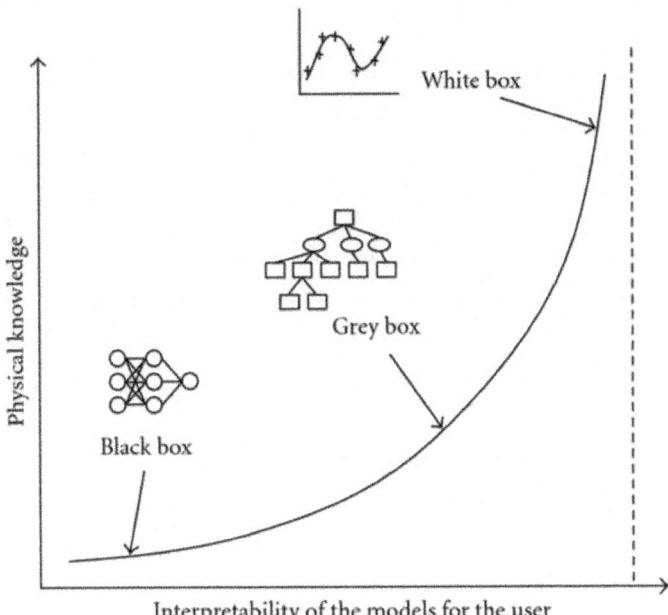

Figure 3: Graphical classification of modeling techniques (adapted from [94]).

ANNs belong to the class of black-box models due to their lack of transparency and the fact that they do not consider nor explain the underlying physical processes explicitly. This is because the knowledge extracted by ANNs is stored in a set of weights that are difficult to interpret properly, and due to the large complexity of the network structure, ANNs fail to give a transparent function that relates the inputs to the corresponding outputs. Consequently, it is difficult to understand the nature of the input-output relationships derived. This issue has been addressed by many researchers with respect to hydrological engineering. For example, Jain et al. [96] examined whether or not the physical processes in a watershed were inherent in a trained ANN rainfall-runoff model. This was carried out by assessing the strengths of the relationships between the distributed components of the ANN model, in terms of the responses from the hidden nodes, and the deterministic components of the hydrological process, computed from a conceptual rainfall runoff model, along with the observed input variables, using correlation coefficients and scatter plots. They concluded that the trained ANN, in fact, captured different components of the physical process and a careful examination of the distributed information contained in the trained ANN can be informative about the nature of the physical processes captured by various components of the ANN model. Sudheer [97] performed perturbation analysis to assess the influence of each individual input variable on the output variable and found it to be an effective means of identifying the underlying physical process inherent in the trained ANN. Olden et al. [89], Sudheer and Jain [98], and Kingston et al. [99] also addressed this issue of model transparency and knowledge extraction.

In the context of geotechnical engineering, Shahin et al. [12] and Shahin and Jaksa [9] expressed the results of the trained ANNs in the form of relatively straightforward equations. This was possible due to the relatively small number of input and output variables, and hidden nodes. Neurofuzzy applications are another means of knowledge extraction that facilitate model transparency. Neurofuzzy networks use the fuzzy logic system to store knowledge acquired from a set of input variables $(x_1, x_2, \ldots, x_n)$ and the corresponding output variable (y) in a set of linguistic fuzzy rules that can be easily interpreted, such as IF (x1 is high AND x_2 is low) THEN (y is high), c = 0.9, where (c = 0.9) is the rule confidence, which indicates the degree to which the above rule has contributed to the output. Examples of such applications in geotechnical engineering include Ni et al. [100], Shahin et al. [16], Gokceoglu et al. [62], Provenzano et al. [19], and Padmini et al. [18].

A recent technique that belongs to the class of grey-box models, and therefore does not suffer from the problem of model transparency and knowledge extraction, is genetic programming (GP). Several researchers (e.g., [34, 50, 101–104]) have recently used the GP technique as an alterative to ANNs in order to obtain greatly simplified formulae for some geotechnical engineering problems. GP is a computing method that attempts to mimic the biological evolution of living organisms. GP makes use of the principles of genetic algorithms (GAs) for parameter optimization in which a population of expressions (or computer programs) for a function F, coded in tree structures of variable size, is generated and executed. The generated expressions are then modified by means of artificial evolution in order to perform a global search to arrive at the best fit mathematical expression for F that solves a certain problem. Additional advantages of GP over ANNs are that the structure and network parameters of ANNs (e.g., number of hidden layers and their number of nodes, transfer functions, learning rate, etc.) should be identified a priori and are usually obtained using adhoc trial-and-error approaches. However, the number and combination of terms, as well as the values of GP modeling parameters, are all evolved automatically during model calibration. However, hybrid approaches can also be used, in which genetic algorithms are used to evolve optimal ANN structures and connection weight values. It should be noted that while white-box models provide maximum transparency, their construction may be difficult to obtain for many geotechnical engineering problems, where the underlying mechanism is not entirely understood.

Model Extrapolation

Model extrapolation is the ability of ANN models to predict well outside the range of the data used for model calibration. It is generally accepted that ANNs perform best when they do not extrapolate beyond the range of the data used for calibration [105–107]. Whilst this is not unlike other models, it is nevertheless an important limitation of ANNs, as it restricts their usefulness and applicability. Extreme value prediction is of particular concern in several areas of civil engineering, such as hydrological engineering, when floods are forecast, as well as in geotechnical engineering when, for example, liquefaction potential and the stability of slopes are assessed. Sudheer et al. [108] highlighted this issue and proposed a methodology, based on the Wilson-Hilferty transformation, for enabling ANN models to predict extreme values with respect to peak river flows. Their methodology yielded superior predictions when compared with those obtained from an ANN model using untransformed data.

Model Uncertainty

Finally, a further limitation of ANNs is that the uncertainty in the predictions generated is seldom quantified [109]. Failure to account for such uncertainty makes it impossible to assess the quality of ANN predictions, which severely limits their efficacy. In an effort to address this, a few researchers have applied Bayesian techniques to ANN training (e.g., [110–113]) in the context of hydrological engineering and Goh et al. [7] with respect to geotechnical engineering. Goh et al. [7] observed that the integration of the Bayesian framework into the back-propagation algorithm enhanced neural network prediction capabilities and provided assessment of the confidence associated with network predictions. Research to date has demonstrated the value of Bayesian neural networks, although further work is needed in the area of geotechnical engineering. Shahin et al. [13, 114] also incorporated uncertainty in the ANN process by developing a series of design charts expressing the reliability of settlement predictions for shallow foundations on cohesionless soils.

DISCUSSION AND CONCLUSIONS

In the field of geotechnical engineering, it is possible to encounter some types of problems that are very complex and not well understood. In this regard, ANNs provide several advantages over more conventional computing techniques. For most traditional mathematical models, the lack of physical understanding is usually supplemented by either simplifying the problem or incorporating several assumptions into the models. Mathematical models also rely on assuming the structure of the model in advance, which may be suboptimal. Consequently, many mathematical models fail to simulate the complex behavior of most geotechnical engineering problems. In contrast, ANNs are a data driven approach in which the model can be trained on input-output data pairs to determine the structure and parameters of the model. In this case, there is no need to either simplify the problem or incorporate any assumptions. Moreover, ANNs can always be updated to obtain better results by presenting new training examples as new data become available. These factors combine to make ANNs a powerful modeling tool in geotechnical engineering.

Despite the success of ANNs in geotechnical engineering and other disciplines, they suffer from some shortcomings that need further attention in the future, including model robustness, transparency and knowledge extraction, extrapolation, and uncertainty. In addition and according to Flood [115], ANNs in civil engineering, including geotechnical engineering, were used mostly as simple vector mapping devices for function modeling of applications that require rarely more than a few tens of neurons without higher-order structuring. Together, improvements in these issues will greatly enhance the usefulness of ANN models and will provide the next generation of applied artificial neural networks with the best way for advancing the field to the next level of sophistication and application. Until such an improvement is achieved, the authors agree with Flood and Kartam [105] that neural networks for the time being might be treated as a complement to conventional computing techniques rather than as an alternative or may be used as a quick check on solutions developed by more time-consuming and in-depth analyses.

REFERENCES

1. M. A. Shahin, M. B. Jaksa, and H. R. Maier, "Artificial neural network applications in geotechnical engineering," Australian Geomechanics, vol. 36, no. 1, pp. 49–62, 2001.

2. I. Ahmad, M. H. El Naggar, and A. N. Khan, "Artificial neural network application to estimate kinematic soil pile interaction response parameters," Soil Dynamics and Earthquake Engineering, vol. 27, no. 9, pp. 892–905, 2007.

3. S. K. Das and P. K. Basudhar, "Undrained lateral load capacity of piles in clay using artificial neural network," Computers and Geotechnics, vol. 33, no. 8, pp. 454–459, 2006.

4. A. M. Hanna, G. Morcous, and M. Helmy, "Efficiency of pile groups installed in cohesionless soil using artificial neural networks," Canadian

Geotechnical Journal, vol. 41, no. 6, pp. 1241–1249, 2004.

5.   H. Ardalan, A. Eslami, and N. Nariman-Zadeh, "Piles shaft capacity from CPT and CPTu data by polynomial neural networks and genetic algorithms," Computers and Geotechnics, vol. 36, no. 4, pp. 616–625, 2009.

6.   M. A. Shahin, "Modelling axial capacity of pile foundations by intelligent computing," in Proceedings of the 2nd BGA International Conference on Foundations (ICOF '08), pp. 283–294, IHS BRE Press, Dundee, Scotland, 2008.

7.   A. T. C. Goh, F. H. Kulhawy, and C. G. Chua, "Bayesian neural network analysis of undrained side resistance of drilled shafts," Journal of Geotechnical & Geoenvironmental Engineering, vol. 131, no. 1, pp. 84–93, 2005.

8.   M. A. Shahin and M. B. Jaksa, "Intelligent computing for predicting axial capacity of drilled shafts," inProceedings of the International Foundation Congress and Equipment Expo (IFCEE '09), ASCE Geotechnical Special Publication, no. 186, pp. 26–33, Orlando, Fla, USA, 2009.·

9.   M. A. Shahin and M. B. Jaksa, "Neural network prediction of pullout capacity of marquee ground anchors," Computers and Geotechnics, vol. 32, no. 3, pp. 153–163, 2005.

10.   M. A. Shahin and M. B. Jaksa, "Pullout capacity of small ground anchors by direct cone penetration test methods and neural networks," Canadian Geotechnical Journal, vol. 43, no. 6, pp. 626–637, 2006. ·

11.   Y.-L. Chen, R. Azzam, and F.-B. Zhang, "The displacement computation and construction pre-control of a foundation pit in Shanghai utilizing FEM and intelligent methods," Geotechnical and Geological Engineering, vol. 24, no. 6, pp. 1781–1801, 2006.

12.   M. A. Shahin, M. B. Jaksa, and H. R. Maier, "Artificial neural network-based settlement prediction formula for shallow foundations on granular soils," Australian Geomechanics, vol. 37, no. 4, pp. 45–52, 2002.

13.   M. A. Shahin, M. B. Jaksa, and H. R. Maier, "Neural network based stochastic design charts for settlement prediction," Canadian Geotechnical Journal, vol. 42, no. 1, pp. 110–120, 2005.

14.   M. A. Shahin, H. R. Maier, and M. B. Jaksa, "Predicting settlement of shallow foundations using neural networks," Journal of Geotechnical & Geoenvironmental Engineering, vol. 128, no. 9, pp. 785–793, 2002. ·

15.   M. A. Shahin, H. R. Maier, and M. B. Jaksa, "Predicting settlement of shallow foundations using neural networks," Journal of Geotechnical &

Geoenvironmental Engineering, vol. 128, no. 9, pp. 785–793, 2002. ·

16. M. A. Shahin, H. R. Maier, and M. B. Jaksa, "Settlement prediction of shallow foundations on granular soils using B-spline neurofuzzy models," Computers and Geotechnics, vol. 30, no. 8, pp. 637–647, 2003. ·

17. Y. L. Kuo, M. B. Jaksa, A. V. Lyamin, and W. S. Kaggwa, "ANN-based model for predicting the bearing capacity of strip footing on multi-layered cohesive soil," Computers and Geotechnics, vol. 36, no. 3, pp. 503–516, 2009.

18. D. Padmini, K. Ilamparuthi, and K. P. Sudheer, "Ultimate bearing capacity prediction of shallow foundations on cohesionless soils using neurofuzzy models," Computers and Geotechnics, vol. 35, no. 1, pp. 33–46, 2008.

19. P. Provenzano, S. Ferlisi, and A. Musso, "Interpretation of a model footing response through an adaptive neural fuzzy inference system," Computers and Geotechnics, vol. 31, no. 3, pp. 251–266, 2004. ·

20. H. Adeli, "Neural networks in civil engineering: 1989–2000," Computer-Aided Civil and Infrastructure Engineering, vol. 16, no. 2, pp. 126–142, 2001.

21. I. A. Basheer, "Stress-strain behavior of geomaterials in loading reversal simulated by time-delay neural networks," Journal of Materials in Civil Engineering, vol. 14, no. 3, pp. 270–273, 2002. ·

22. Q. Fu, Y. M. A. Hashash, S. Jung, and J. Ghaboussi, "Integration of laboratory testing and constitutive modeling of soils," Computers and Geotechnics, vol. 34, no. 5, pp. 330–345, 2007.·

23. G. Habibagahi and A. Bamdad, "A neural network framework for mechanical behavior of unsaturated soils," Canadian Geotechnical Journal, vol. 40, no. 3, pp. 684–693, 2003.

24. Y. M. A. Hashash, S. Jung, and J. Ghaboussi, "Numerical implementation of a neural network based material model in finite element analysis," International Journal for Numerical Methods in Engineering, vol. 59, no. 7, pp. 989–1005, 2004.

25. M. Lefik and B. A. Schrefler, "Artificial neural network as an incremental non-linear constitutive model for a finite element code," Computer Methods in Applied Mechanics and Engineering, vol. 192, no. 28–30, pp. 3265–3283, 2003.

26. Y. M. Najjar and C. Huang, "Simulating the stress-strain behavior of Georgia kaolin via recurrent neuronet approach," Computers and Geotechnics, vol. 34, no. 5, pp. 346–361, 2007.·

27. M. A. Shahin and B. Indraratna, "Modeling the mechanical behavior of

railway ballast using artificial neural networks," Canadian Geotechnical Journal, vol. 43, no. 11, pp. 1144–1152, 2006.

28. M. Banimahd, S. S. Yasrobi, and P. K. Woodward, "Artificial neural network for stress-strain behavior of sandy soils: knowledge based verification," Computers and Geotechnics, vol. 32, no. 5, pp. 377–386, 2005.

29. W. Gao, X. T. Feng, and Y. R. Zheng, "Identification of a constitutive model for geo-materials using a new intelligent bionics algorithm," International Journal of Rock Mechanics and Mining Sciences, vol. 41, supplement 1, pp. 454–459, 2004.

30. Y. Yang and M. S. Rosenbaum, "The artificial neural network as a tool for assessing geotechnical properties," Geotechnical and Geological Engineering, vol. 20, no. 2, pp. 149–168, 2002.

31. S. Çelik and Ö. Tan, "Determination of preconsolidation pressure with artificial neural network," Civil Engineering and Environmental Science, vol. 22, no. 4, pp. 217–231, 2005.

32. S. J. Lee, S. R. Lee, and Y. S. Kim, "An approach to estimate unsaturated shear strength using artificial neural network and hyperbolic formulation," Computers and Geotechnics, vol. 30, no. 6, pp. 489–503, 2003.

33. P. U. Kurup and N. K. Dudani, "Neural networks for profiling stress history of clays from PCPT data,"Journal of Geotechnical & Geoenvironmental Engineering, vol. 128, no. 7, pp. 569–579, 2002.

34. B. S. Narendra, P. V. Sivapullaiah, S. Suresh, and S. N. Omkar, "Prediction of unconfined compressive strength of soft grounds using computational intelligence techniques: a comparative study," Computers and Geotechnics, vol. 33, no. 3, pp. 196–208, 2006.

35. A. Baykasoglu, H. Güllü, H. Çanakçi, and L. Özbakir, "Prediction of compressive and tensile strength of limestone via genetic programming," Expert Systems with Applications, vol. 35, no. 1-2, pp. 111–123, 2008.

36. W. Y. Byeon, S. R. Lee, and Y. S. Kim, "Application of flat DMT and ANN to Korean soft clay deposits for reliable estimation of undrained shear strength," International Journal of Offshore and Polar Engineering, vol. 16, no. 1, pp. 73–80, 2006.

37. A. Kaya, "Residual and fully softened strength evaluation of soils using artificial neural networks,"Geotechnical and Geological Engineering, vol. 27, no. 2, pp. 281–288, 2009.

38. Y. Erzin, "Artificial neural networks approach for swell pressure versus soil suction behaviour,"Canadian Geotechnical Journal, vol. 44, no. 10,

pp. 1215–1223, 2007.

39. I. Ashayeri and S. Yasrebi, "Free-swell and swelling pressure of unsaturated compacted clays; experiments and neural networks modeling," Geotechnical and Geological Engineering, vol. 27, no. 1, pp. 137–153, 2009.

40. S. K. Das and P. K. Basudhar, "Prediction of coefficient of lateral earth pressure using artificial neural networks," The Electronic Journal of Geotechnical Engineering, vol. 10, pp. 1–5, 2005.

41. S. K. Sinha and M. C. Wang, "Artificial neural network prediction models for soil compaction and permeability," Geotechnical and Geological Engineering, vol. 26, no. 1, pp. 47–64, 2008.

42. A. H. Abdel-Rahman, "Predicting compaction of cohesionless soils using ANN," Ground Improvement, vol. 161, no. 1, pp. 3–8, 2008.

43. P. U. Kurup and E. P. Griffin, "Prediction of soil composition from CPT data using general regression neural network," Journal of Computing in Civil Engineering, vol. 20, no. 4, pp. 281–289, 2006.

44. B. Bhattacharya and D. P. Solomatine, "Machine learning in soil classification," Neural Networks, vol. 19, no. 2, pp. 186–195, 2006.

45. M. P. Romo and S. R. García, "Neurofuzzy mapping of CPT values into solid dynamic properties," Soil Dynamics and Earthquake Engineering, vol. 23, no. 6, pp. 473–482, 2003.

46. S. R. García, M. P. Romo, and J. Figueroa-Nazuno, "Soil dynamic properties determination: a neurofuzzy system approach," Control and Intelligent Systems, vol. 34, no. 1, pp. 1–11, 2006.

47. M. H. Baziar and A. Ghorbani, "Evaluation of lateral spreading using artificial neural networks," Soil Dynamics and Earthquake Engineering, vol. 25, no. 1, pp. 1–9, 2005.

48. A. T. C. Goh, "Probabilistic neural network for evaluating seismic liquefaction potential," Canadian Geotechnical Journal, vol. 39, no. 1, pp. 219–232, 2002.

49. A. M. Hanna, D. Ural, and G. Saygili, "Neural network model for liquefaction potential in soil deposits using Turkey and Taiwan earthquake data," Soil Dynamics and Earthquake Engineering, vol. 27, no. 6, pp. 521–540, 2007.

50. A. A. Javadi, M. Rezania, and M. M. Nezhad, "Evaluation of liquefaction induced lateral displacements using genetic programming," Computers

and Geotechnics, vol. 33, no. 4-5, pp. 222–233, 2006.

51. K. Young-Su and K. Byung-Tak, "Use of artificial neural networks in the prediction of liquefaction resistance of sands," Journal of Geotechnical & Geoenvironmental Engineering, vol. 132, no. 11, pp. 1502–1504, 2006.

52. A. M. Hanna, D. Ural, and G. Saygili, "Evaluation of liquefaction potential of soil deposits using artificial neural networks," Engineering Computations, vol. 24, no. 1, pp. 5–16, 2007.·

53. M. S. Rahman and J. Wang, "Fuzzy neural network models for liquefaction prediction," Soil Dynamics and Earthquake Engineering, vol. 22, no. 8, pp. 685–694, 2002.

54. C. H. Juang, H. Yuan, D.-H. Lee, and P.-S. Lin, "Simplified cone penetration test-based method for evaluating liquefaction resistance of soils," Journal of Geotechnical & Geoenvironmental Engineering, vol. 129, no. 1, pp. 66–80, 2003.

55. S. S. H. Khozaghi and A. J. A.-Z. Choobbasti, "Predicting of liquefaction potential in soils using artificial neural networks," Electronic Journal of Geotechnical Engineering, vol. 12C, 2007.

56. G. T. C. Kung, E. C. L. Hsiao, M. Schuster, and C. H. Juang, "A neural network approach to estimating deflection of diaphragm walls caused by excavation in clays," Computers and Geotechnics, vol. 34, no. 5, pp. 385–396, 2007.

57. Y.-S. Kim and B.-T. Kim, "Prediction of relative crest settlement of concrete-faced rockfill dams analyzed using an artificial neural network model," Computers and Geotechnics, vol. 35, no. 3, pp. 313–322, 2008.

58. Y. Yu, B. Zhang, and H. Yuan, "An intelligent displacement back-analysis method for earth-rockfill dams," Computers and Geotechnics, vol. 34, no. 6, pp. 423–434, 2007.

59. Y. Lu, "Underground blast induced ground shock and its modelling using artificial neural network,"Computers and Geotechnics, vol. 32, no. 3, pp. 164–178, 2005.

60. T. N. Singh and V. Singh, "An intelligent approach to prediction and control ground vibration in mines," Geotechnical and Geological Engineering, vol. 23, no. 3, pp. 249–262, 2005.·

61. J. Q. Shang, W. Ding, R. K. Rowe, and L. Josic, "Detecting heavy metal contamination in soil using complex permittivity and artificial neural networks," Canadian Geotechnical Journal, vol. 41, no. 6, pp. 1054–1067, 2004.

62. C. Gokceoglu, E. Yesilnacar, H. Sonmez, and A. Kayabasi, "A neuro-

fuzzy model for modulus of deformation of jointed rock masses," Computers and Geotechnics, vol. 31, no. 5, pp. 375–383, 2004. ·

63. T. N. Singh, A. K. Verma, V. Singh, and A. Sahu, "Slake durability study of shaly rock and its predictions," Environmental Geology, vol. 47, no. 2, pp. 246–253, 2005.

64. S. Ma, L.-H. Cao, and H.-Y. Li, "The improved neutral network and its application for valuing rock mass mechanical parameter," Journal of Coal Science and Engineering, vol. 12, no. 1, pp. 21–24, 2006. ·

65. T. N. Singh, A. K. Verma, and P. K. Sharma, "A neuro-genetic approach for prediction of time dependent deformational characteristic of rock and its sensitivity analysis," Geotechnical and Geological Engineering, vol. 25, no. 4, pp. 395–407, 2007.

66. V. B. Maji and T. G. Sitharam, "Prediction of elastic modulus of jointed rock mass using artificial neural networks," Geotechnical and Geological Engineering, vol. 26, no. 4, pp. 443–452, 2008. ·

67. T. G. Sitharam, P. Samui, and P. Anbazhagan, "Spatial variability of rock depth in Bangalore using geostatistical, neural network and support vector machine models," Geotechnical and Geological Engineering, vol. 26, no. 5, pp. 503–517, 2008.

68. N. Caglar and H. Arman, "The applicability of neural networks in the determination of soil profiles,"Bulletin of Engineering Geology and the Environment, vol. 66, no. 3, pp. 295–301, 2007.

69. C. Yoo and J.-M. Kim, "Tunneling performance prediction using an integrated GIS and neural network," Computers and Geotechnics, vol. 34, no. 1, pp. 19–30, 2007.

70. A. G. Benardos and D. C. Kaliampakos, "Modelling TBM performance with artificial neural networks,"Tunnelling and Underground Space Technology, vol. 19, no. 6, pp. 597–605, 2004.·

71. K. Neaupane and S. Achet, "Some applications of a backpropagation neural network in geo-engineering," Environmental Geology, vol. 45, no. 4, pp. 567–575, 2004.

72. K. M. Neaupane and N. R. Adhikari, "Prediction of tunneling-induced ground movement with the multi-layer perceptron," Tunnelling and Underground Space Technology, vol. 21, no. 2, pp. 151–159, 2006.

73. Y.-L. Chen, R. Azzam, T. M. Fernandez-Steeger, and L. Li, "Studies on construction pre-control of a connection aisle between two neighbouring tunnels in Shanghai by means of 3D FEM, neural networks and fuzzy logic," Geotechnical and Geological Engineering, vol. 27, no. 1, pp.

155–167, 2009. ·

74. A. Alimoradi, A. Moradzadeh, R. Naderi, M. Z. Salehi, and A. Etemadi, "Prediction of geological hazardous zones in front of a tunnel face using TSP-203 and artificial neural networks," Tunnelling and Underground Space Technology, vol. 23, no. 6, pp. 711–717, 2008.

75. M. D. Ferentinou and M. G. Sakellariou, "Computational intelligence tools for the prediction of slope performance," Computers and Geotechnics, vol. 34, no. 5, pp. 362–384, 2007.

76. A. T. C. Goh and F. H. Kulhawy, "Neural network approach to model the limit state surface for reliability analysis," Canadian Geotechnical Journal, vol. 40, no. 6, pp. 1235–1244, 2003.

77. F. Mayoraz and L. Vulliet, "Neural networks for slope movement prediction," The International Journal of Geomechanics, vol. 2, no. 2, pp. 153–173, 2002.

78. D. P. Kanungo, M. K. Arora, S. Sarkar, and R. P. Gupta, "A comparative study of conventional, ANN black box, fuzzy and combined neural and fuzzy weighting procedures for landslide susceptibility zonation in Darjeeling Himalayas," Engineering Geology, vol. 85, no. 3-4, pp. 347–366, 2006.

79. H. B. Wang and K. Sassa, "Rainfall-induced landslide hazard assessment using artificial neural networks," Earth Surface Processes and Landforms, vol. 31, no. 2, pp. 235–247, 2006.·

80. Y. M. A. Hashash, C. Marulanda, J. Ghaboussi, and S. Jung, "Systematic update of a deep excavation model using field performance data," Computers and Geotechnics, vol. 30, no. 6, pp. 477–488, 2003. ·

81. L. V. Fausett, Fundamentals Neural Networks: Architecture, Algorithms, and Applications, Prentice-Hall, Englewood Cliffs, NJ, USA, 1994.

82. J. M. Zurada, Introduction to Artificial Neural Systems, West Publishing, Saint Paul, Minn, USA, 1992.

83. H. R. Maier and G. C. Dandy, "Applications of artificial neural networks to forecasting of surface water quality variables: issues, applications and challenges," in Artificial Neural Networks in Hydrology, R. S. Govindaraju and A. R. Rao, Eds., pp. 287–309, Kluwer Academic Publishers, Dordrecht, The Netherlands, 2000.

84. M. A. Shahin, H. R. Maier, and M. B. Jaksa, "Data division for developing neural networks applied to geotechnical engineering," Journal of Computing in Civil Engineering, vol. 18, no. 2, pp. 105–114, 2004. ·

85. M. A. Shahin, M. B. Jaksa, and H. R. Maier, "State of the art of artificial

neural networks in geotechnical engineering," Electronic Journal of Geotechnical Engineering, vol. 8, pp. 1–26, 2008.

86. G. B. Kingston, H. R. Maier, and M. F. Lambert, "Calibration and validation of neural networks to ensure physically plausible hydrological modeling," Journal of Hydrology, vol. 314, no. 1-4, pp. 158–176, 2005.

87. M. A. Shahin, H. R. Maier, and M. B. Jaksa, "Investigation into the robustness of artificial neural network models for a case study in civil engineering," in Proceedings of the International Congress on Modelling and Simulation (MODSIM '05), pp. 79–83, Melbourne, Australia, 2005.

88. G. D. Garson, "Interpreting neural-network connection weights," AI Expert, vol. 6, no. 7, pp. 47–51, 1991.

89. J. D. Olden, M. K. Joy, and R. G. Death, "An accurate comparison of methods for quantifying variable importance in artificial neural networks using simulated data," Ecological Modelling, vol. 178, no. 3-4, pp. 389–397, 2004.

90. M. Pal, "Support vector machines-based modelling of seismic liquefaction potential," International Journal for Numerical and Analytical Methods in Geomechanics, vol. 30, no. 10, pp. 983–996, 2006.

91. A. T. C. Goh and S. H. Goh, "Support vector machines: their use in geotechnical engineering as illustrated using seismic liquefaction data," Computers and Geotechnics, vol. 34, no. 5, pp. 410–421, 2007.

92. H.-B. Zhao, "Slope reliability analysis using a support vector machine," Computers and Geotechnics, vol. 35, no. 3, pp. 459–467, 2008.

93. P. Samui, "Prediction of friction capacity of driven piles in clay using the support vector machine,"Canadian Geotechnical Journal, vol. 45, no. 2, pp. 288–295, 2008.

94. O. Giustolisi, A. Doglioni, D. A. Savic, and B. W. Webb, "A multi-model approach to analysis of environmental phenomena," Environmental Modelling & Software, vol. 22, no. 5, pp. 674–682, 2007. ·

95. O. Giustolisi, "Using genetic programming to determine Chezy resistance coefficient in corrugated channels," Journal of Hydroinformatics, vol. 3, no. 6, pp. 157–173, 2004.

96. A. Jain, K. P. Sudheer, and S. Srinivasulu, "Identification of physical processes inherent in artificial neural network rainfall runoff models," Hydrological Processes, vol. 18, no. 3, pp. 571–581, 2004.

97. K. P. Sudheer, "Knowledge extraction from trained neural network river flow models," Journal of Hydrologic Engineering, vol. 10, no. 4, pp. 264–269, 2005.

98. K. P. Sudheer and A. Jain, "Explaining the internal behaviour of artificial neural network river flow models," Hydrological Processes, vol. 18, no. 4, pp. 833–844, 2004.

99. G. B. Kingston, H. R. Maier, and M. F. Lambert, "A probabilistic method for assisting knowledge extraction from artificial neural networks used for hydrological prediction," Mathematical and Computer Modelling, vol. 44, no. 5-6, pp. 499–512, 2006.

100. S. H. Ni, P. C. Lu, and C. H. Juang, "A fuzzy neural network approach to evaluation of slope failure potential," Computer-Aided Civil and Infrastructure Engineering, vol. 11, no. 1, pp. 59–66, 1996.

101. X.-T. Feng, B.-R. Chen, C. Yang, H. Zhou, and X. Ding, "Identification of visco-elastic models for rocks using genetic programming coupled with the modified particle swarm optimization algorithm,"International Journal of Rock Mechanics and Mining Sciences, vol. 43, no. 5, pp. 789–801, 2006.

102. A. Johari, G. Habibagahi, and A. Ghahramani, "Prediction of soil-water characteristic curve using genetic programming," Journal of Geotechnical & Geoenvironmental Engineering, vol. 132, no. 5, pp. 661–665, 2006.

103. M. Rezania and A. Javadi, "A new genetic programming model for predicting settlement of shallow foundations," Canadian Geotechnical Journal, vol. 44, no. 12, pp. 1462–1472, 2007.

104. I. Alkroosh, M. A. Shahin, and H. R. Nikraz, "Modelling axial capacity of bored piles using genetic programming technique," in Proceedings of the 3rd International Geo-Chiangmai Conference, pp. 113–120, Chiangmai, Thailand, 2008.

105. I. Flood and N. Kartam, "Neural networks in civil engineering. I: principles and understanding,"Journal of Computing in Civil Engineering, vol. 8, no. 2, pp. 131–148, 1994. ·

106. A. W. Minns and M. J. Hall, "Artificial neural networks as rainfall-runoff models," Hydrological Sciences Journal, vol. 41, no. 3, pp. 399–417, 1996.

107. A. S. Tokar and P. A. Johnson, "Rainfall-runoff modeling using artificial neural networks," Journal of Hydrologic Engineering, vol. 4, no. 3, pp. 232–239, 1999.

108. K. P. Sudheer, P. C. Nayak, and K. S. Ramasastri, "Improving peak flow estimates in artificial neural network river flow models," Hydrological Processes, vol. 17, no. 3, pp. 677–686, 2003.·

109. H. R. Maier and G. C. Dandy, "Neural networks for the prediction and forecasting of water resources variables: a review of modelling issues

and applications," Environmental Modelling & Software, vol. 15, no. 1, pp. 101–124, 2000.

110. W. L. Buntine and A. S. Weigend, "Bayesian back-propagation," Complex Systems, vol. 5, pp. 603–643, 1991.

111. G. B. Kingston, M. F. Lambert, and H. R. Maier, "Bayesian parameter estimation applied to artificial neural networks used for hydrological modelling," Water Resources Research, vol. 41, Article ID W12409, 2005.

112. G. MacKay, "A practical Bayesian framework for backpropagation networks," Neural Computation, vol. 4, pp. 448–472, 1992.

113. G. B. Kingston, H. R. Maier, and M. F. Lambert, "Bayesian model selection applied to artificial neural networks used for water resources modeling," Water Resources Research, vol. 44, no. 4, Article ID W04419, 2008.

114. M. A. Shahin, M. B. Jaksa, and H. R. Maier, "Stochastic simulation of settlement of shallow foundations based on a deterministic neural network model," in Proceedings of the International Congress on Modelling and Simulation (MODSIM '05), pp. 73–78, Melbourne, Australia, 2005.

115. I. Flood, "Towards the next generation of artificial neural networks for civil engineering," Advanced Engineering Informatics, vol. 22, no. 1, pp. 4–14, 2008.

Chapter 6

THE LOW COMPACTION GRADING TECHNIQUE ON STEEP RECLAIMED SLOPES: SOIL CHARACTERIZATION AND STATIC SLOPE STABILITY

Isaac A. Jeldes[1], Eric C. Drumm[2], John S. Schwartz[1]

[1]Department of Civil and Environmental Engineering, University of Tennessee, Knoxville, TN 37996, USA

[2]Department of Biosystems Engineering and Soil Science, University of Tennessee, Institute of Agriculture, Knoxville, TN 37996, USA

ABSTRACT

Since the Surface Mining and Control Reclamation Act of 1977, US coal mining companies have been required by law to restore the approximate ground contours that existed prior to mining. To ensure mass stability and limit erosion, the reclaimed materials have traditionally been placed with significant compaction energy. The Forest Reclamation Approach (FRA) is a relatively new approach that has been successfully used to facilitate the fast establishment of native healthy forests. The FRA method specifies the use of low compaction energy in the top 1.2–1.5 m of the contour, which may be in conflict with general considerations for mechanical slope stability. Although successful for reforestation, the stability of FRA slopes has not been fully investigated and a rational stability method has not been identified. Further, a mechanics-based analysis is limited due to the significant amount of oversize particles which makes the sampling and measurement of soil strength properties difficult. To investigate the stability of steep FRA slopes (steeper than 20°), three reclaimed coal mining sites in the Appalachian region of East Tennessee were investigated. The stability was evaluated by several methods to identify the predominant failure modes. The infinite slope method, coupled with the estimation of the shear strength from field observations, was shown to provide a rational means to evaluate the stability of FRA slopes. The analysis results suggest that the low compaction of the surface materials may not compromise the long-term stability for the sites and material properties investigated.

INTRODUCTION

A reclamation method that employs minimally compacted spoils to enhance native forest growth, known as the Forest Reclamation Approach (FRA) is currently being promoted by the US Office of Surface Mining (OSM) (Angel et al. 2007; Sweigard et al. 2007b). Since the Surface Mining and Control Reclamation Act of 1977 (SMCRA), coal companies in the USA have been required by law to restore the land to its pre-mined contours (USDoI 1977). Reclamation activities have traditionally incorporated compaction procedures to augment the strength of the reclaimed material and ensure stability of the restored slopes. However, while compaction is important for strength and erosion resistance, it diminishes soil porosity which restricts root penetration and reduces water infiltration with negative impacts on tree survival and grass reestablishment (Angel et al. 2007; Sweigard et al. 2007b). FRA employs a low compaction effort in the uppermost 1.2–1.5 m. The low-compaction grading technique has proven to be successful in encouraging tree growth, and demonstrates the potential for establishing healthy native forests on reclaimed mine lands (Angel et al. 2007; Barton et al. 2007). However, with the exception of Torbert and Burger (1994), most of these demonstrations were conducted on relatively flat lying terrain where stability issues were negligible. The stability of steep FRA slopes, defined as steeper than 20° by the USDoI (2009), and the possible modes of failure have not been investigated and a rational stability analysis method has not been suggested.

Slope stability analysis requires the knowledge of soil properties in terms of density and strength; characteristics not easily determined for reclaimed mine spoil due to the significant amount of oversize material (> 0.3 m). The in situ density of soils consisting of large rock particles can be difficult to measure, which makes it difficult to quantify and awkward to provide proper construction quality control. Sweigard et al. (2007a) have suggested correlations between dry bulk density and shovel penetration; though practical for reforestation efforts they are not appropriate for the evaluation of slope stability. Furthermore, because of the difficulties associated with sampling and testing due to the oversize particles, the shear strength properties are not typically measured in laboratory or field tests. For mine reclamation projects, the design is typically completed well in advance of mining activities and usually based on experience using assumed or traditional regional soil properties (Bell et al. 1989). Naturally, there is uncertainty associated with this practice, especially if low compaction is employed on steep reclaimed slopes. For example, in Kentucky the majority of slope failures in abandoned mine lands have occurred via translational and rotational failure mechanisms through the loose material placed prior to the SMCRA (Iannacchione and Vallejo 1995). The lack of proper compaction is a known cause of failure in constructed slopes, with the stability becoming worse under intense rainstorms (Chen et al. 2004).The failure of Sau Mau Ping slopes in Hong Kong is one dramatic example of the danger associated with poorly compacted slopes and lack of proper engineering design (Abramson 1996; Hong Kong Geotechnical Engineering Office 2007).

The objectives of this paper are to (1) characterize the geotechnical properties of low compacted spoils on steep slopes constructed according to the FRA, and (2) investigate the likely failure mechanisms associated to steep slopes reclaimed using the FRA. This is accomplished using three reclaimed field sites at which the material characteristics are evaluated, and the results will be used to suggest a practical method to estimate the shear strength and evaluate the stability of slopes constructed using the low compaction grading technique.

METHODS

Location of Field Sites

To investigate the potential effects on stability resulting from the implementation of the low compaction grading technique, three steep FRA slopes were studied. The three sites, referred to here by the name of the initial coal operator (Premium, National and Mountainside), are located in northeastern Tennessee, with Premium located in Anderson County, National in Campbell County and Mountainside located in Claiborne County (Figure. 1). Each of the mine operators played an instrumental role in the development of the study sites. Each site was divided into four different plots which while not discussed here, were instrumented in order to concurrently investigate the runoff hydrology and sediment yield on the FRA slopes (Hoomehr et al. 2013). Figure. 2 shows the National site during construction of the study plots.

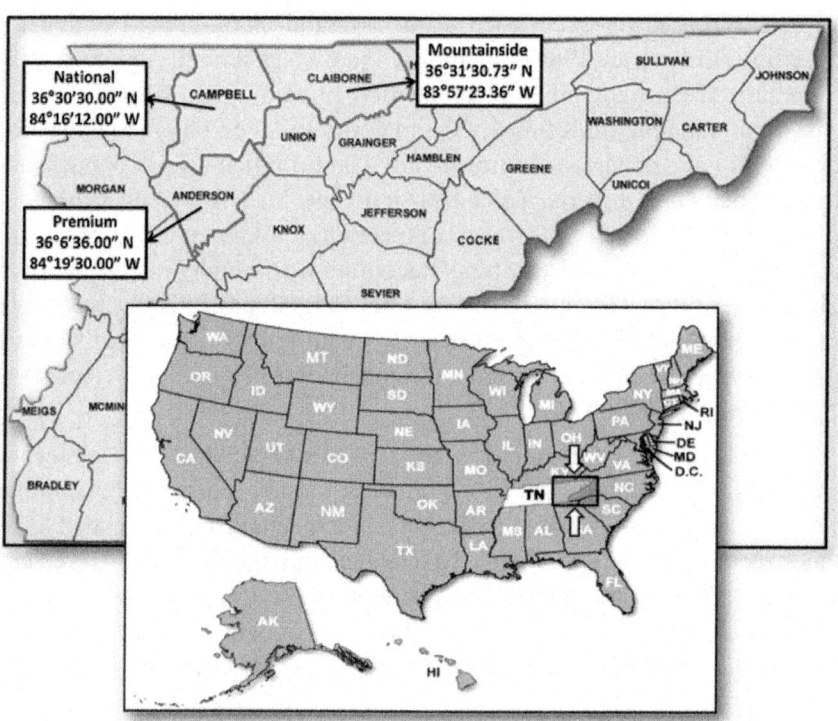

Figure. 1: Location of field sites in northeastern Tennessee, referred to as Premium, National, and Mountainside

Figure. 2: National site after the FRA reclamation process and during the construction of the study plots.

Site Construction and Reclamation Process

At each of the three sites in this study, the construction procedure followed the contour *haulback method*(Sweigard and Kumar 2010), where a ramp is constructed on the contour bench and spoil is hauled up the ramp and dumped over the edge. The sequence of the construction process can be divided into four major steps (Sweigard et al. 2007b) depicted schematically in Figure. 3: (a) placement and compaction of the materials for the primary backfill core using traditional practices, (b) dumping of the soil that will constitute the loose surface layer (1.2–1.5 m thick), (c) grading of the loose soil layer with the lightest equipment available using the fewest passes possible, and (d) reforestation. The three research sites presented a very rough soil surface after the final grading, which is consistent with the FRA recommendations for successful reforestation (Sweigard et al. 2007b). However, because the final layer at all three sites often included boulder-sized material, significant depressions and large rocks were left on the surface of the slope which is a deviation of Sweigard's recommendations for an ideal finished surface.

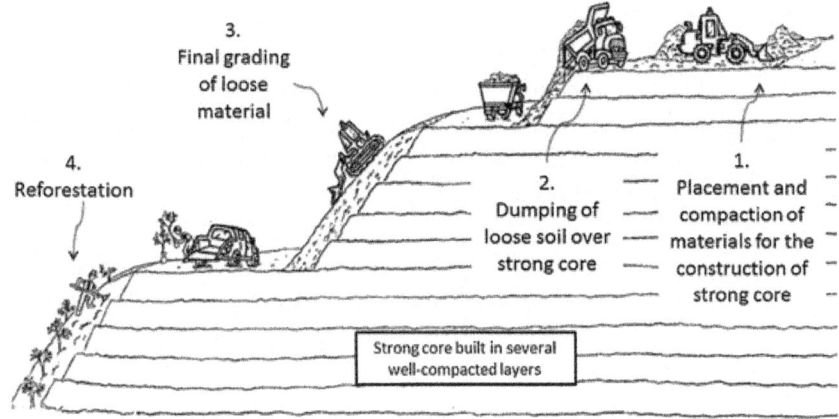

Figure. 3: Depiction of the reclamation process according to FRA.

Geotechnical Characterization

The investigation proceeded with the characterization of the research sites and the analysis of their mechanical stability. The field characterization of the mine spoil included: (a) determination of the site geometry; (b) particle size analysis, index tests, and classification of the materials; (c) determination of unit weight; and (d) estimation of the Mohr–Coulomb (M–C) shear strength parameters.

Geometry

The geometric characteristics of the research sites were determined via a series of Trimble™ total station topographical surveys. The purpose of these surveys was not only to obtain a more accurate estimation of the slope angles, but also to gather data to geo-reference the instrumented sites allowing an investigation of the spatial variation of material properties.

Particle Size Analyses, Index Tests, and Classification of Reclaimed Materials

Four soil samples of 0.02 m^3 each were randomly taken across the slope at each site for particle size analyses, index tests and classification. All samples were collected from a depth of at least 30 cm below the slope surface to (a) avoid samples with fewer fines due to erosion armoring and (b) avoid surficial soils affected by changes in fabric due to weathering. Particle size analysis (grain size distribution and hydrometer) and Atterberg limits tests were conducted in general accordance to ASTM D422-07 and ASTM D4318–10 respectively, and classification of the materials in general accordance to the Unified Soil Classification System (USCS), ASTM D2487-10. Visual inspection suggested that the materials may contain a large amount of agglomerated fines in the form of large particles, and therefore, traditional dry particle size analysis would indicate a larger amount of coarse material than really exists. We investigated this issue at all three sites by allowing soil samples to soak in water for 14 days. We found very few aggregated fines in the Premium and National soils, but significant aggregated fines at Mountainside. For this reason, a wet preparation of the Mountainside samples was conducted in general accordance to ASTM D2217-85 before conducting the particle size analysis and index tests. The amount of oversize material (> 0.3 m) was estimated on a surface basis. This was accomplished by dividing the plot into multiple squares of 1 m side length; a photograph of each square was used to estimate the percentage of oversize particles per surface area.

Unit Weight

An extensive data collection of the unit weight of the loose surface layer at each of the three sites was conducted using a Troxler 3411-B Nuclear Density Gage (NDG), in general accordance to the ASTM D6938-10. The measurements were obtained shortly after each slope was constructed. During the data collection, periodic calibration of the NDG device was conducted at the beginning and middle of each work day, employing the calibration block provided by the manufacturer. A randomized systematic sampling technique (Sweigard et al.

2007c) was used at one plot per site to reduce data tendency or bias in the measurements. The plot was divided into multiple squares of 3 m side length, where single measurements of bulk dry unit weight $_d$, bulk wet unit weight $_T$ and moisture content w were taken at random locations inside the squares (Figure. 4). The soil surface was cautiously carved to obtain a planar surface before placing the nuclear gauge device, avoiding air interchange between the gauge base and the soil surface. Then, a hole perpendicular to the slope surface was driven inside each sub-area and the source rod inserted to obtain the readings at 300 mm (which is the maximum length of the source rod of the Nuclear Density Gauge). From previous observations of the materials it was concluded that the amount of hydrocarbons present at each site was very small and unlikely to affect the NDG readings.

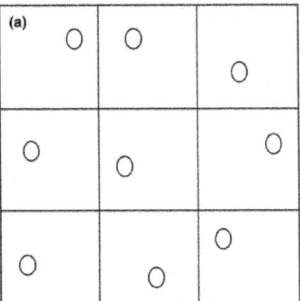

Figure. 4: Randomized systematic sampling technique; **a** area of interest subdivided into small sub-areas (Sweigard et al. 2007c), where *small circles* represent the random measurement locations; **b** application of this technique for NDG measurements at the National site.

Shear Strength Parameters

Slope stability analysis for long-term conditions assumes that positive excess pore water pressure dissipates during the construction or loading period, and thus, requires the estimation of the drained or effective shear strength parameters. The shear strength of soil is typically described using the M–C yield or failure criterion:

$$\tau_n = c + \sigma_n \tan \phi$$

(1)

where σ_n and τ_n are the normal and shear stresses acting on the failure plane within the soil body, and ϕ and c are the internal friction angle and cohesion. Outside of mine reclamation, the shear strength material properties ϕ and c are often measured in laboratory or obtained through correlations with in situ tests. However, due to the large particles present in mine spoils, traditional tests are very difficult to conduct. The current practice in mine reclamation is usually based on experience or assumed values of ϕ and c. As discussed earlier, the FRA technique consists on having a 1.2–1.5 m of loose soil at the uppermost part of the contour, above a well compacted and stable core. Thus, this low density/ low strength zone should be evaluated for stability. Since the angle of repose is the "steepest stable slope for loose packed granular material and represents the angle of internal friction at its loosest state" (Holtz and Kovacs 1981), it is suggested as a good representation of the internal friction angle of loose soil layer in a FRA slope. The use of observed angles of repose offers the additional advantage that the overall strength of the mass, including the contribution of the oversize particles, is captured. Angles of repose were obtained by observing the placement of spoils piles, and measuring the angle at which they hold in place. Because these tests were conducted during reclamation activities, we measured the angle via photographs of the fresh piles using a hand held level, for safety reasons. The level helped us to ensure that the picture was taken parallel to the horizon. With a photo editing software the angles of the piles were measured (Figure. 5). Regarding the strength properties of the stronger core material, stability analysis will later show that the most critical condition for stability is insensitive to the selected strength values of the core.

Figure. 5: Typical field determination of the angle of repose at National Site (White et al. 2009). Camera placed on a level and photo taken of material in loose state. Note the large number of oversize (> 0.3 m) particles in the reclaimed material.

On the other hand, short-term stability analyses (undrained soil conditions) are often conducted to characterize the behavior of the soil during and immediately after construction, where the loading occurs much faster than the rate of dissipation of positive excess pore water pressure. However, it is assumed that the coarse mine spoil material will not develop significant excess positive pore water pressure under typical loadings (Duncan and Wright 2005); thus, a drained response is expected. This assumption is supported by the grain size distribution of the materials and relatively high void ratios obtained for each site which are reported in the results section. Furthermore, short-term stability is not considered to be important in the reclamation of mine slopes, since any short-term failure would have minimum consequences and would be repaired during construction or routine maintenance.

Static Long-Term Slope Stability Analyses

The analyses of the mechanical stability focused on long-term analyses of the primary failure modes that are likely to be experienced by FRA slopes: (a) shallow or local failure modes within the loose surface layer and (b) global or deep rotational failure modes of the overall soil mass. Limit Equilibrium Methods (LEM) and the Finite Element Method (FEM) were employed in the analyses assuming 2-D plain strain conditions. LEM analyses were computed using Slide 6.0 (Rocscience Inc. 2011a) with 10,000 critical surfaces analyzed. The FEM analyses were computed using Phase2 (Rocscience Inc. 2011b) employing an elastic perfectly plastic stress–strain behavior and the M–C yield criterion with a non-associated flow rule (zero dilatancy angle) to avoid over-prediction of dilation and failure load for purely frictional materials (Griffiths and Lane 1999). Since the estimation of the factor of safety (FS) has been shown to be not significantly affected by the value of the elastic constants E (Young's modulus) and v (Poisson's ratio) used in the FEM solution (Griffiths and Lane1999; Cheng et al. 2007), nominal values of $E = 10^5$ kPa and $v = 0.3$ were assumed here for the surface and core materials. The model was created with 6-noded triangular elements, with a maximum of 500 iterations solved by Gaussian elimination. Because the geometry and material properties from all sites are reasonably similar, those from Mountainside ($\beta = 28°$ from the horizontal and $H = 21$ m height) will be used here to explore the various failure modes. These values are representative of all three sites and many steep slopes in the southern Appalachian coal fields. The thickness of the low-strength layer (z) was assumed to be 1.5 m.

Shallow Stability Within the Low Strength Surface Layer

A shallow failure mode was investigated by assuming that the core was significantly stronger than the surface layer. The methods used were: (a) LEM restricting the analyses to the shallow surface layer via the non-circular Janbu's method and a search block feature, (b) FEM and the shear strength reduction method (Griffiths and Lane 1999; Cheng et al. 2007), and (c) the infinite slope equation for cohesionless soils without seepage $(FS = \tan\phi/\tan\beta)$ and with seepage $(FS \approx 0.5\tan\phi/\tan\beta)$. The infinite slope equation idealizes the surface as an infinite plane with the failure mechanism running parallel to the surface (Skempton and Delory 1957) and it is appropriate when the ratio of depth to length of the sliding surface is small. The geometry of the reclaimed mine slopes constructed according to the FRA is ideally suited for investigation by the infinite slope method.

Deep Rotational Stability of the Overall Soil Mass

Because the strength of the compacted core was not known, the deep rotational failure mode was investigated by performing a series of analyses where the strength of the loose surface material was held constant ($\phi = 38°$, $c = 0$) while the internal friction angle of the core was increased. The analyses started with a homogeneous slope with the properties of the weak, loose surface layer analyzed via LEM (the circular Simplified Bishop's method) and FEM. Then, the shear strength of the core was increased such that the ratio $\tan\phi_{core}/\tan\phi_{loose}$ was equal to 1.1, 1.2, 1.3, and 1. (ϕ_{core} core is the friction angle of the core material and ϕ_{loose} is the friction angle of the loose surface layer).

RESULTS AND DISCUSSION

Geotechnical Characterization of Research Sites

Geometry

The geometric information of the three sites obtained from the topographical survey is summarized in Table 1. Preliminary information of the slopes angles via a Suunto PM-5/360PC mechanical inclinometer reported angles between 26° and 30° at Premium site, 20° and 22° at National site, and 28° and 29° at Mountainside site (White et al. 2009). These angles, collected shortly after the end of the reclamation process, coincide with the topographic information gathered 15 months later, and suggest no changes in the slope morphology and no slope failures during the study period.

Table 1: Average slope length, width and inclination angle for the four plots at the premium, national and mountainside sites

Site	Average slope angle, β (°)	Average slope length (m)	Average slope width (m)	
			Top	Bottom
Premium	28	32.2	28.1	25.0
National	20	48.4	22.4	25.4
Mountainside	28	45.4	23.6	23.1

Particle Size Analysis, Index Tests, and Classification of Reclaimed Materials

Results from the soil analyses are reported in Table 2. The grain size distribution was conducted on material smaller than 51 mm (2 in. sieve), while Atterberg limits were determined on material smaller than 0.42 mm (No. 40 sieve). According to the USCS, for all the research sites the material classify as clayey gravel (GC) with the exception of one plot at the Premium site that classifies as poorly graded clayey gravel (GP-GC) due to slightly less material finer than the number 200 sieve. Regarding oversize particles, it was estimated that material larger than 300 mm occupies 0–25 % of 1 m^2 at Premium site, 0–10 % of 1 m^2 at National site, and 5–40 % of 1 m^2 at Mountainside site.

Table 2: Mean values of Liquid Limit, Plastic Index, soil texture and soil classification (USCS)

Sites	Gravel particles 51–4.75 mm (%)	Sand particles 4.75–0.075 mm (%)	Fines <0.075 mm (%)	Clay particle <2 μm (%)	Liquid limit (LL)	Plastic index (PI)	Soil classification (USCS)
Premium	59	28	13	6	29	13	GC to GP-GC
National	52	28	20	10	27	14	GC
Mountainside	37	22	41	19	32	15	GC

Unit Weight

Results of the unit weight measurements were as follows: the maximum and minimum measured $_d$ were 18.8 and 13.0 kN/m^3 at Premium, 21.4 and 14.6 kN/m^3 at National, and 22.8 and 14.9 kN/m^3 at Mountainside. Complementary results from statistical analyses are presented in Table 3. Field measures of density at a similar mine site in Kentucky (Sweigard et al. 2011) indicated similar variations as those found at Premium and National. The largest standard deviations (SD) were observed at Mountainside, which is consistent with the largest range of unit weights, and the largest amount of fines and observed number of oversize particles. At all sites, the spatial variation of unit weight reflects the large range of particle sizes in these reclaimed materials. A single unit weight measurement has a limited ability to represent the state of body forces acting on the complete FRA slope, and thus, mean unit weights with the probable upper and lower bounds (confidence and tolerance intervals) are desired for the mine material characterization. While confidence intervals (C.I.) provide an upper and lower bound of the true mean found at the constructed sites, tolerance intervals (T.I.) provide information of the probable future range of unit weights that each site will have on average. In any case, as discussed later, the determination of the unit weight for static long-term conditions may be of minor concern, but necessary for static unsaturated and seismic stability analyses.

Table 3: Means, standard deviations, 95 % confidence intervals (C.I.), and 90 % tolerance intervals (T.I.) (80 % coverage) for wet and dry unit weights for premium, national and mountainside sites

Sites	Unit weight	Mean (kN/ m^3)	SD (kN/ m^3)	95 % C.I. for the mean		90 %/0.8 T.I. for the mean	
				Lower (kN/m^3)	Upper (kN/m^3)	Lower (kN/ m^3)	Upper (kN/m^3)
Premium	Dry	16.2	1.3	15.8	16.5	14.2	18.1
	Wet	18.5	1.3	18.2	18.8	16.6	20.4
National	Dry	18.5	1.0	18.3	18.7	17.2	19.9
	Wet	20.3	1.0	20.1	20.5	18.9	21.7
Mountainside	Dry	18.6	2.2	18.1	19.1	15.5	21.7
	Wet	20.4	2.2	19.9	20.9	17.2	23.6

Water and sand cone replacement tests were previously conducted to determine bulk unit weights at random locations on the four plots at each site (White et al. 2009). A comparison of the results indicates that the NDG device gives on average about 25 % higher average unit weights than replacement methods at Premium site, 14 % at National site and 21 % at Mountainside. Since both replacement methods involved the removal of small samples, they did not take into account the effects of large rock fragments that are randomly embedded in the loose surface soil layer. On the other hand, the NDG calculates unit weights based on the velocity travel of gamma rays between the source and the detector, and any denser material that appears on the travel path will be counted in the measurement. In this regard, the collection of a sufficient amount of NDG readings will better represent the wide range of in-place density and provides a more representative average unit weight for stress analyses. Replacement methods may be preferable for the calculation of void ratio and soil porosity due to better representation of the soil matrix. The average void ratio of the loose surface layer calculated via replacement methods was 1.0 at Premium, 0.6 at National and 0.7 at Mountainside. The largest void ratio was calculated for the soils found at Premium, which is consistent with the lowest NDG unit weight measured. Overall, relatively large void ratios were obtained for all three sites, which is consistent with the FRA requirements for healthy tree growth.

Shear Strength Parameters

The angle of repose of the looser soil layer was found to range between 37° and 39° at Premium and Mountainside, and between 36° and 38° at National site. Zero cohesion is usually employed for long-term analysis on coarse granular soils (Lambe and Whitman 1969; Holtz and Kovacs 1981) and normally consolidated fine soils (Skempton 1964), and would be appropriate for reclaimed materials receiving minimum compaction effort. While even a small amount of compaction will increase the density and strength of the soil, the angle of repose is a conservative estimate of the friction angle. Similar values of ϕ and c for loose spoils in the Appalachian region were reported by Sweigard et al. (2011), while similar values for reclaimed spoils outside the Appalachian were found in the literature (Ulusay et al. 1995; Stormont and Farfan 2005; Kasmer and Ulusay 2006; Gutierrez et al. 2008; Sweigard et al. 2011) as summarized in Table 4.

Table 4: Summary of internal friction angle and cohesion for reclaimed mine materials

Author	Origin of material tested	Type of test	Sample dimensions (mm)	Internal friction angle $\phi(°)$	Cohesionc (kN/m²)
Ulusay et al. (1995)	Limestone, claystone and marl (Turkey)	In situ SPT test	N/A	31–38	N/A
Ulusay et al. (1995).	Limestone, claystone and marl (Turkey)	Direct shear test	N/A	34 (peak) 33 (residual)	12 (peak) 9 (residual)
Ulusay et al. (1995)	Limestone, claystone and marl (Turkey)	Triaxial (CD) test	Diameter = 191 Height = 382	23-35	0–10
Stormont and Farfan (2005)	N/A (San Juan, Colorado)	Direct shear test (large laboratory box)	Length = 762 Width = 762 Height = 457	37	5
Gutierrez et al. (2008)	N/A (Northern New Mexico)	Direct shear test	Length = 51 Width = 51 Height = N/A	42–47 (peak) 37–41 (residual)	0
Kasmer and Ulusay (2006)	Limestone and mar (Turkey)	Direct shear test	N/A	31–34 (peak) 24–33 (residual)	18–34 (peak) 6–10 (residual)
Sweigard et al. (2011)	Sandstone and shale (Pike County, Kentucky)	Triaxial (CU) test	N/A	37	0

FRA re-search sites (this study)	Sandstone and shale (Northeast Tennessee)	Angle of repose	N/A	38	0

Static Long-Term Slope Stability Analyses

Shallow Stability Within the Low Strength Surface Layer

Results from the limit equilibrium, finite element, and infinite slope analyses are summarized in Table 5. From a practical perspective, all the analyses yielded very similar FS's (approximately 1.47), implying that the shear strength along the most critical slip surface is about 47 % greater than that required to maintain equilibrium in the long-term. For all cases, the most critical failure mechanism is shallow and is consistent with the assumed failure mechanism in the infinite slope method. Figure. 6 shows the FEM model of the shallow failure mode with a section of the slope enlarged (the 1.5 m thick surface layer is small with respect to the size of the model and may not be clearly distinguished in the full model). It also shows nodal displacement vectors. Larger strains are observed at the interface of the weak surface and core materials, with the displacement vectors acting parallel to the surface suggesting a planar failure mechanism. The obtained long-term FS's are valid for drained conditions in the absence of seepage forces due to transient flow. However, since the occurrence of downslope water flow through the complete thickness of the loose layer is highly unlikely, this condition represents a lower bound or worst case value for the stability of FRA, and would reduce the FS by a factor of 2.

Table 5: FS obtained for long-term static stability focused on low strength surface layer

Analysis method	Assumptions	FS	Critical failure mode
(a) Limit equilibrium	Rigid core and search block—non-linear Janbu's method with 10,000 critical surfaces analyzed	1.48	Shallow planar failure surface
(b) Finite element method	Core much stronger than loose surface layer, shear strength reduction method to determine FS, with 500 iterations solved by Gaussian elimination	1.47	Shallow planar failure surface

(c) Analytical	Infinite slope equation (no seepage)	1.47	Shallow planar failure surface

Slope   stability   analysis   results   for   generic   slope ($\beta = 28°$, $H = 21$ m, $\phi = 38°$, $c = 0$, $\gamma_T = 20.4$ kN/m³)

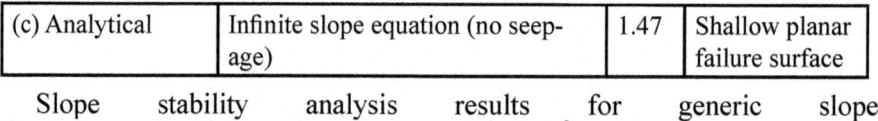

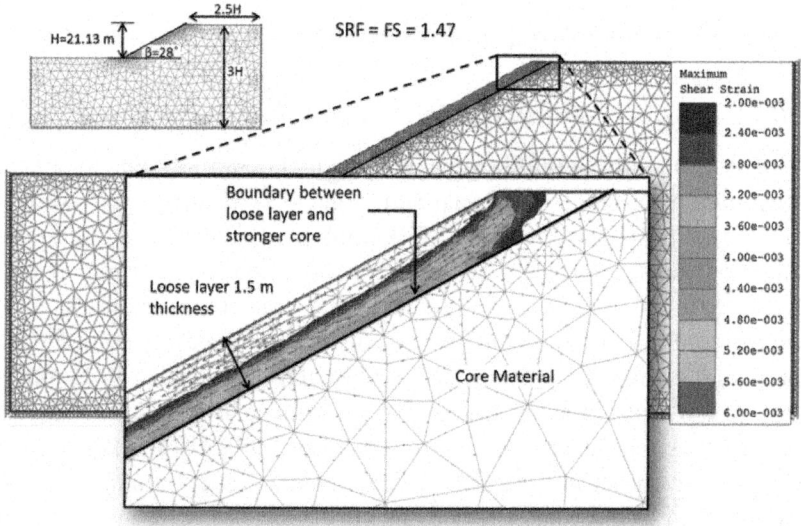

Figure.6: Shear strains and nodal displacements obtained from the FEM analysis assuming a very strong core. *Upper left corner* illustrates the geometric dimensions employed for LEM and FEM analyses

Deep Rotational Stability of the Overall Soil Mass

The results from LEM and FEM analyses of the deep failure mode of the homogeneous slope yield a FS = 1.48, which is consistent with that obtained from the shallow stability analyses. Figure. 7 shows the results of a FEM analysis when the core strength was 30 % stronger than the loose layer $\tan \phi_{core} / \tan \phi_{loose} = 1.3$). Here two possible failure mechanisms were observed in the form of shear bands; a deeper mechanism through the core material with a FS = 1.94, and a shallow mechanism with the lowest FS = 1.48 and highest shear strains at the interface of the materials. The FS of the shallow mechanism is equal to those obtained from the shallow analyses above. A similar trend is observed for cases when $\tan \phi_{core} / \tan \phi_{loose} = 1.1, 1.2,$ and 1.4 (Figure. 8)As the strength of the core increases, the FS of the deeper mechanism increases; however, the lowest FS is found to be constant with a consistent shallow failure mode and dependent only on the strength level of the loose surface layer. Additional analyses at angles of inclination of 20° and 35° yielded similar results and confirm that the critical failure mode is a surface failure.

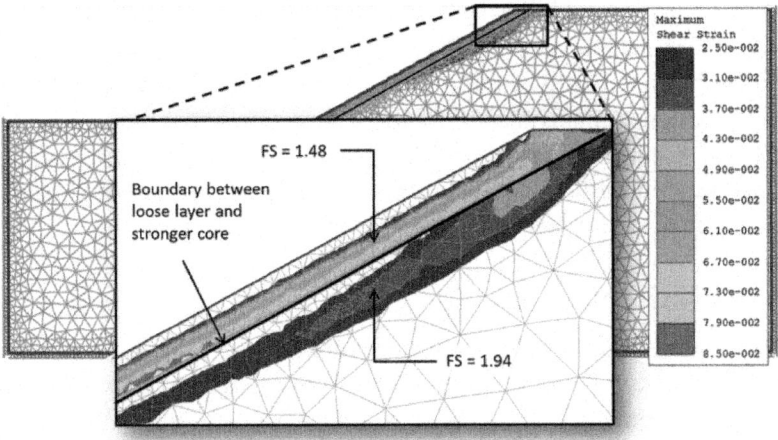

Figure.7: Failure mechanisms and FS's obtained from FEM stability analysis for $\tan \phi_{core} / \tan \phi_{loose} = 1.3.$ The FS = 1.94 shown for the deeper failure mechanism was obtained when the strength reduction factor (SRF) search was restricted to be outside the zone where the shallow mechanism occurred. Shear strains showed for SRF = 2.02 to emphasize failure mode

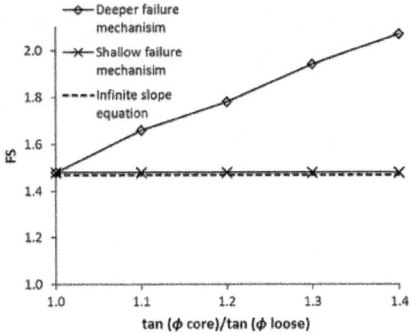

Figure.8:FEM analyses of the global stability for various values of $\tan \phi_{core} / \tan \phi_{loose}$ and the infinite slope equation.

These results are consistent with the observation that the failure mechanisms through the loose surface layer will govern and the determination of the strength parameters of the stronger dense core are not important for FRA slope design. Furthermore, since the infinite slope method adequately approximates the shallow failure mode, and accurately predicts the FS, it can be taken as a simple and reliable method to evaluate the performance of FRA slopes and more sophisticated computer analyses are not necessary for most applications. The use of the infinite slope method also simplifies the field characterization of materials and disregards the unit weight determination, because it only requires ϕ_{loose} and β for long-term conditions. The simplicity of the method is appropriate for design of reclaimed mine slopes which are typically designed in advance of mineral extraction with assumed overburden properties. Accordingly, the lowest FS for drained or long-term conditions at each instrumented site using the infinite slope equation is approximately 1.47 for Premium, 2.07 for National and 1.47 for Mountainside.

CONCLUSIONS

- Characterization and stability evaluation of three FRA slopes in northeastern Tennessee with inclinations as high as 28° were conducted. A large number of oversize particles were found in the reclaimed materials. In general, the material finer than 51 mm classified as clayey gravels with the average Plasticity Index (PI) ranging from 13 to 15, suggesting that the physical characteristic of the soils are similar across the three research sites.

- Unit weights determined using a Nuclear Density Gage were found to be higher than those determined by replacement methods, yet vary significantly across the study plots. NDG measures are preferred for stability analyses because they better capture the effect of oversize particles on the in situ state of stresses of FRA slopes. It also allows more measurements to characterize the wide range of in-place density. Tolerance intervals were constructed to reflect the probable future range of unit weights that each site will have on average.

- The analysis of several potential modes of failure suggests that the governing failure mode is shallow and contained within the weak, loose surface layer. The determination of the strength parameters of the core is not important for FRA slope design.

- Because the infinite slope method adequately approximates the shallow failure mode and accurately predicts the FS, it may be an appropriate method to evaluate the performance of FRA slopes and more sophisticated analyses are not necessary for most applications. Since the unit weight of the material is not considered in the infinite slope expression, field measurements of the highly variable unit weight are not required for long-term analyses.

- The angle of repose was suggested to be a conservative estimate of the internal friction angle and it is consistent with the loose nature of the

FRA material. This provides a means to quantify the friction angle of the mine spoil, which has been traditionally assumed based on experience.

- The shear strength along the most critical slip surface, for the typical FRA slope investigated, is at least 47 % greater than that required to maintain static equilibrium in the long-term. In case that the entire loose surface zone becomes saturated with downslope seepage and no infiltration into the core, the FS is reduced by a factor of 2, suggesting that the slope would be unstable. However, these conditions are very unlikely and provide a lower limit to the factor of safety.

- The likely conditions would suggest that the FRA has no negative impact on slope stability, and the benefits of faster forest establishment in terms of reduced erosion and sediment delivery make the FRA very attractive for future reclamation work.

ACKNOWLEDGMENTS

This research was conducted as part of a project funded by the US Office of Surface Mines, Applied Science Program Grant CA No. S08AP12822, 2008. The authors gratefully appreciate the support and guidance provided by OSM staff David Lane, P. E., and Vic Davis, and the field assistance provided by Patrick White, Wesley Wright, Esteban Zamudio, Mitch Groothuis and Siavash Hoomehr.

REFERENCES

1. Abramson LW (1996) Slope stability and stabilization methods. Wiley, New York

2. Angel PC, Barton C, Warner R, Agouridis C, Taylor T, Hall S (2007) Hydrologic characteristics, tree growth, and natural regeneration on three loose-graded surface mine spoil types in Kentucky. Paper presented at the Mid-Atlantic Stream Restoration Conference, Cumberland, MD, 7–9 November

3. Barton C, Agouridis C, Warner R, Bidelspach D, Angel P, Jennings G, Marchant J, Osborne R (2007) Recreating a headwater stream system on a head-of-hollow fill. Paper presented at the Mid-Atlantic Stream Restoration Conference, Cumberland, MD, 7–9 November

4. Bell JC, Daniels WL, Zipper CE (1989) The practice of "approximate original contour" in the central Appalachians. I. Slope stability and erosion potential. Landsc Urban Plan 18(2):127–138.

5.   Chen H, Lee CF, Law KT (2004) Causative mechanisms of rainfall-induced fill slope failures. J Geotech Geoenviron Eng 130(6):593–602.

6. Cheng YM, Lansivaara T, Wei WB (2007) Two-dimensional slope stability analysis by limit equilibrium and strength reduction methods. Comput Geotech 34(3):137–150.

7. Duncan JM, Wright SG (2005) Soil strength and slope stability. Wiley, Hoboken, NJ

8. Griffiths DV, Lane PA (1999) Slope stability analysis by finite elements. Géotechnique 49(3):387–403

9. Gutierrez LAF, Viterbo VC, McLemore VT, Aimone-Martin CT (2008) Geotechnical and geomechanical characterization of the Goathill north rock pile at the Questa Molybdenum Mine, New Mexico, USA. In: Fourie A (ed) First International Seminar on the Management of Rock Dumps, Stockpiles and Heap Leach Pads, Perth, Australia, 5–6 March 2008. The Australian Centre for Geomechanics, pp 19–32

10. Holtz RD, Kovacs WD (1981) An introduction to geotechnical engineering. Prentice-Hall, Englewood Cliffs, NJ

11. Hong Kong Geotechnical Engineering Office (2007) Engineering geological practice in Hong Kong Geo Publication No. 1/2007

12. Hoomehr S, Schwartz J, Yoder D, Drumm E, Wright W (2013) Curve numbers for low-compaction steep-sloped reclaimed mine lands in the southern Appalachians. J Hydrol Eng (in press).

13. Iannacchione AT, Vallejo LE (1995) Factors affecting the slope stability of Kentucky abandoned mine lands. In: Daemen J, Schultz R (eds) Proceedings of the 35th US Symposium on Rock Mechanics, Reno, NV, 5–7 June 1995. A. A. Balkema, Rotterdam, Netherlands, pp 837–842

14. Kasmer O, Ulusay R (2006) Stability of spoil piles at two coal mines in Turkey: geotechnical characterization and design considerations. Environ Eng Geosci 12(4):337–352

15. Lambe TW, Whitman RV (1969) Soil mechanics. Series in soil engineering. Wiley, New York

16. Rocscience Inc. (2011) Slide 6.0. 2D Limit equilibrium slope stability analysis, version 6.0 edn. Rocscience Inc., Toronto, Canada

17. Rocscience Inc. (2011) Phase2 7.0. Finite element analysis for excavations and slopes, version 8.0 edn. Rocscience Inc., Toronto, Canada

18. Skempton AW (1964) Long-term stability of clay slopes. Géotechnique 14(2):77–101CrossRef

19. Skempton AW, Delory FA (1957) Stability of natural slopes in London clay. In: 4th International Conference on Soil Mechanics & Foundation Engineering, London. pp 378–381

20. Stormont JC, Farfan E (2005) Stability evaluation of a mine waste pile. Environ Eng Geosci 11(1):43–52

21. Sweigard RJ, Kumar D (2010) Filed investigation of best practices for steep slope mine reclamation employing the forestry reclamation approach. In: Proceedings of the joint 27th Annual American Society of Mining and Reclamation and 4th Annual Appalachian Regional Reforestation Initiative, Pittsburgh, PA, 5–11 June 2010. American Society of Mining and Reclamation

22. Sweigard R, Burger J, Graves D, Zipper C, Barton C, Skousen J, Angel P (2007a) Loosening compacted soils on mined sites. Forest Reclamation Advisory No4

23. Sweigard RJ, Badaker V, Hunt K (2007b) Development of a field procedure to evaluate the reforestation potential of reclaimed surface-mined land. Department of Mining Engineering, University of Kentucky, Lexington, KY

24. Sweigard R, Burger J, Zipper C, Skousen J, Barton C, Angel P (2007c) Low compaction grading to enhance reforestation success on coal surface mines. Forest Reclamation Advisory No. 3

25. Sweigard R, Hunt K, Kumar D (2011) Field investigation of best practices for steep-slope mine reclamation employing the Forestry Reclamation Approach, Final Report. US Department of the Interior, Office of Surface Mining Reclamation and Enforcement, Denver, CO

26. Torbert JL, Burger JA (1994) Influence of grading intensity on ground cover establishment, erosion, and tree establishment on steep slopes. In: International Land Reclamation and Mine Drainage Conference and the Third International Conference on the Abatement of Acidic Drainage, Pittsburgh, PA, 24–29. American Society of Mining and Reclamation, pp 226–231

27. Ulusay R, Arikan F, Yoleri MF, Caglan D (1995) Engineering geological characterization of coal mine waste material and an evaluation in the context of back-analysis of spoil pile instabilities in a strip mine, SW Turkey. Eng Geol 40(1–2):77–101

28. USDoI (1977) Surface mining reclamation and enforcement provisions. US Department of the Interior. Federal Register. http:// www. gpo. gov

29. USDoI (2009) 30 CFR 701.5—permanent regulatory program.US Department of the Interior. http:// www. gpo. gov. 5 July 2012

30. White PH, Drumm EC, Schwartz JS, Johnson AM (2009) Geotechnical characterization of steep slopes on. Reclaimed Mine Lands in East Tennessee. Paper presented at the 2009 ASABE Annual International Meeting, Reno, NV, 21–24 June

Chapter 7

NATURAL VARIABILITY OF SHEAR STRENGTH IN A GRANITE RESIDUAL SOIL FROM PORTO

Luı́s Pinheiro Branco[1], Antonio Topa Gomes[2],Anto´nio Silva Cardoso[2],Carla Santos Pereira[3]

[1]L. Pinheiro Branco (&) AdFGeo - Consultores de Geotecnia, Rua Ferna˜o Lopes, 157 - 48 Esq, Porto 4150-318, Portugal

[2]A. Topa Gomes A. Silva Cardoso Department of Civil Engineering, Faculty of Engineering, University of Porto, Rua Dr. Roberto Frias, Porto 4200-465, Portugal

[3]C. Santos Pereira Portucalense University Infante D. Henrique, Rua Dr. Anto´nio Bernardino de Almeida, 541, Porto 4200-072, Portugal

ABSTRACT

The renewal imposed by the Eurocodes regarding the methodologies of safety evaluation requires a statistical analysis of the variability of ground geotechnical parameters. However, the studies published in the reviewed literature do not cover the typical materials from the northeast region of Portugal—residual soils from granite—to which a strong heterogeneity is associated. Hence, a statistical characterization of the natural variability of a granite residual soil from Porto has been made through a significant amount of experimental tests, focusing on its geomechanical properties. In order to provide a database for probabilistic analysis of problems involving this type of soils, an appropriate statistical law has been used to model its variability, which has been quantified by means of coefficients of variation and scales of fluctuation.

INTRODUCTION

The concept of safety and its evaluation has experienced a remarkable evolution over the last few years. However, the determination of a global factor of safety is still widely used in the design of geotechnical structures, which creates additional difficulties in understanding the influence in design resulting from the uncertainties in the different parameters. Therefore, following the renewal imposed by Eurocodes, deterministic methodologies will tendentiously be replaced by more rational approaches, such as semi-probabilistic methods—for example, the partial coefficients method—and probabilistic methods based on the reliability theory.

An extensive characterization of the variability of geotechnical parameters must be carried out to enable the transition to be fulfilled. This characterization should be as objective as possible, avoiding ambiguities provided by qualitative descriptions, as suggested by Kulhawy and Phoon (1999) who recommend a statistical analysis including the coefficient of variation and the scale of fluctuation. Although the literature provides benchmarks for sands and clays, quantitative studies devoted to the variability of residual soils are scarce and their singular characteristics require specific treatment.

Accordingly, Pinheiro Branco (2011) performed a significant amount of direct shear tests in order to characterize the natural variability of a granite residual soil from Porto, namely its shear strength but also some physical properties. The test results were subjected to a statistical treatment focused on the coefficient of variation and the scale of fluctuation and the main conclusions are discussed in this paper.

RESIDUAL SOILS VARIABILITY

Uncertainties in Geotechnics

The main difference between geotechnics and other fields of civil engineering has to do with the fact that geotechnical problems involve natural materials, namely soils and rocks, whose properties depend on natural processes which humans cannot control. Consequently, site geotechnical characterization involves inferences to be carried out from limited data and dealing with different sources of uncertainty (see Figure. 1).

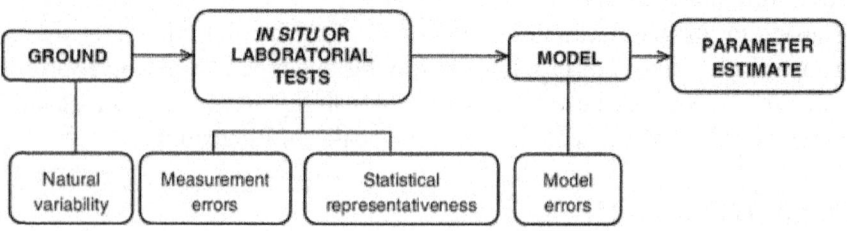

Figure.1: Sources of uncertainty in geotechnics (Kulhawy and Phoon 1999, adapted).

As shown in Figure. 1, there are three main sources of uncertainty when estimating ground parameters—natural variability, measurement errors and model errors. As this paper deals with residual soils, which present a strong heterogeneity, natural variability takes additional importance and is going to be the principal topic addressed.

The Particular Case of Residual Soils

Residual soils derive from weathering of underlying parent rocks, having a wide particle size distribution and a bonded structure, where coarse grains (in the case of granite residual soils, usually minerals of quartz) are bonded by fragile clayey bridges (Viana da Fonseca et al. 1997).

In contrast to sedimentary deposits, which generally present horizontal stratification, residual soils profiles are particularly random. Indeed, there might be sites in which there is a mass of residual soil surrounded by intact rock and other sites where boulders can be found within a thick layer of residual soil (Viana da Fonseca et al.2010).

So, as residual soils genesis depends on weathering factors which are not constant in place, a variability analysis should also take into account its spatial components, both vertical and horizontal.

Natural Variability Quantification

Coefficient of Variation and Scale of Fluctuation

Kulhawy and Phoon (1999) suggest the use of two quantitative attributes to analyze ground natural variability, namely the coefficient of variation and the scale of fluctuation.

The coefficient of variation is a normalized measure of data dispersion which, for a sample of size n (y_1, y_2, ... y_n), can be estimated for an arbitrary variable Y using eq. (1) (Curto and Pinto 2009).

$$\widehat{cv}_y = \frac{\sqrt{\frac{\sum_{i=1}^{n}(y_i - \widehat{\mu}_y)^2}{n-1}}}{\widehat{\mu}_y}$$

(1)

Where $\widehat{\mu}_y$ is the sample estimate of the mean (that is, the arithmetic mean of n data values which constitute the sample of Y).

Regarding the scale of fluctuation, consider, by way of example, the unsupported slope of residual soil illustrated in Figure. 2 which is on the verge of sliding along a relic discontinuity of its parent rock. Figureure 2 also presents a graphical representation of the variation of the friction angle, $\emptyset$, along the weak strength plane, as well as its mean value.

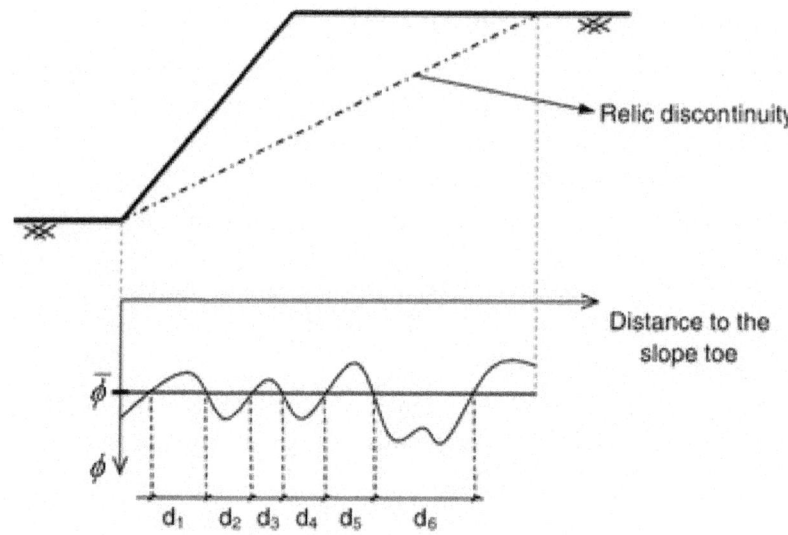

Figure.2: Cross section of a slope in residual soil and friction angle along a relic discontinuity.

As shown in Figure. 2, the friction angle along the sliding surface is not constant. So, in order to take into account spatial fluctuations around the mean value, it is opportune to introduce the concept of scale of fluctuation, δ. The scale of fluctuation of a given property measures the distance over which the ground properties present values of the same order of magnitude, at least with respect to the tendency defined by the mean value, indicating the existence of some correlation between the properties of adjacent points (Vanmarcke 1977). Its value can be determined roughly using eq. (2) (Kulhawy and Phoon 1999).

$$\delta \simeq 0.8 \times \bar{d} \tag{2}$$

in which $\bar{d}$ is the average distance between intersections of the curve representing the real value of a given property and its average curve (see Figure. 2).

In practice, the scale of fluctuation is crucial to understand the behavior of potentially unstable masses of soil. Indeed, soils behave as highly hyperstatic structures; therefore, if a particular loading leads to failure of the most fragile areas, there can be a redistribution of shear stresses to adjacent areas with greater strength. However, this load redistribution can only take place before failure if the surface on which shear strength is mobilized is large enough to ensure that the overall soil behavior is governed by it average resistance (Silva Cardoso and Matos Fernandes 2001). That is, the scale of fluctuation represents a point of reference used to determine if the failure surface is large enough so that shear stresses can really be redistributed. In order to take that internal redistribution capacity into account in calculations, Silva Cardoso and Matos Fernandes (2001) suggest the use of a corrected coefficient of variation, cv_{real}, defined as:

$$cv_{real} = cv \times \prod_{i=1}^{n} \left(\frac{\delta_i}{L_i}\right)^{1/2}$$

(3)

where the dimensions of a potentially unstable mass of soil are defined by L_i, while δ_i are the corresponding scales of fluctuation; note that i refers to a particular spatial direction out of the n along which the fluctuations of the property in analysis are considered.

Typical Statistical Parameters and Scales of Fluctuation in Soils

Table 1 summarizes common coefficients of variation related to some physical and mechanical properties of sandy and clayey soils, constituting the main references with regard to the characterization of the natural variability of soils.

Table 1: Typical coefficients of variation in soils

Property	Soil Type	cv (%)	Reference
Unit weigth	Sands and clays	3–7	Duncan (2000)
Effective friction angle	Sands	3–12	Duncan (2000)
		5–15	Baecher and Christian (2003)
	Clays	12–56	Baecher and Christian (2003)
Effective cohesion	Sands and clays	10–70	Shahin and Cheung (2011)
		20–40	Forrest and Orr (2010)
Voids ratio	Sands and clays	7–42	Baecher and Christian (2003, adapted)

Concerning the spatial component of variability, the parameters characterized in terms of scale of fluctuation are mainly the ones that can be evaluated from in situ tests that provide a continuous record of ground properties. The values indicated for undrained shear strength obtained from vane-tests and for CPT tip resistance are presented in Table 2, resulting that horizontal scales of fluctuation tend to be one order of magnitude greater than the vertical ones.

Table 2: Typical scales of fluctuation in soils (Kulhawy and Phoon 1999, adapted)

Property	Soil type	Direction	δ (m)
Undrained shear strength	Clays	Vertical	0.8–6.2
		Horizontal	46.0–60.0
Tip Resistance (CPT)	Sands and clays	Vertical	0.1–2.2
		Horizontal	3.0–80.0

The variability of most geotechnical parameters can be modeled with normal distributions. However, when data values are low, spread out and cannot be negative, as it happens with cohesion, using normal distributions may not be appropriate. In fact, Forrest and Orr (2010) suggest instead the use of a lognormal distribution to model the variability of cohesion, since it only assumes non-negative values and gives a greater weight to lower values.

NATURAL VARIABILITY OF A GRANITE RESIDUAL SOIL FROM PORTO

Samples of residual soil were collected in Porto, Portugal (see Figure. 3)— geographical coordinates in the UTM Datum WGS84: 41.17330° (latitude) and −8.60195° (longitude).

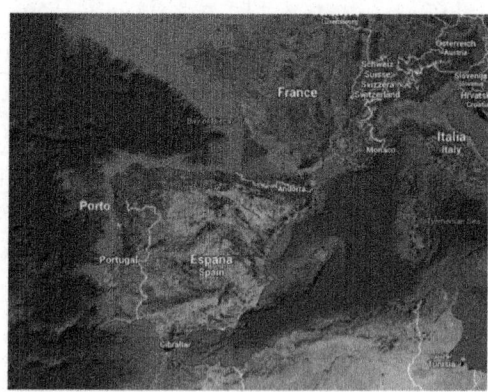

Figure.3: Map showing the geographical location of Porto (Google Maps 2014).

Geologically, it is a residual soil from granite, with a W5 weathering state, presenting a stained aspect, where it is possible to differentiate a whitish stain with completely weathered feldspars from a yellowish stain, which has more sand and presents oxidized biotites (see Figure. 4). It was found that this heterogeneity was not only random in plan view, but also in depth (see Figure. 4b).

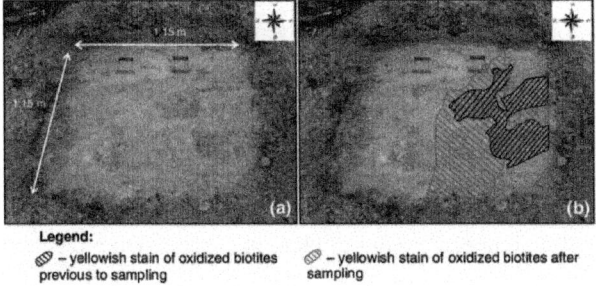

Legend:
⊘ – yellowish stain of oxidized biotites   ⊘ – yellowish stain of oxidized biotites after
previous to sampling   sampling

Figure.4: Detail of study area after topsoil removing: **(a)** previous to sampling **(b)** evolution of the yellowish stain of oxidized biotites after sampling.

The natural variability of shear strength in the study area of approximately 1.15×1.15 m^2, presented in Figure. 4, was characterized through 40 standard direct shear tests. Samples were grouped together in 10 sets of four samples each, as schematized in Figure. 5, for which peak and constant volume shear strength were evaluated in drained conditions by tests conducted under normal stresses of 25, 50, 75 and 100 kPa.

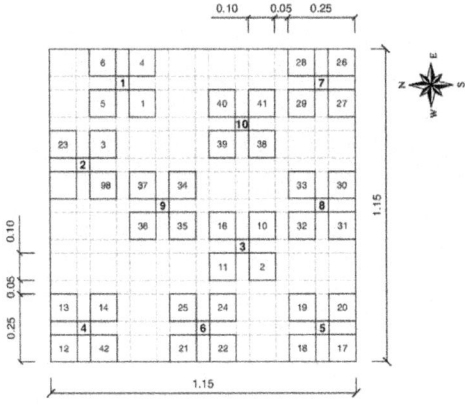

Figure.5: Study area scheme with the ten sets of samples.

All procedures related to direct shear tests, including in situ sampling, preservation, transportation and preparation of samples in the laboratory, were carried out according to a standard test method—ASTM D3080-04. Succinctly, and highlighting some important aspects, the procedure was the following:

- undisturbed samples were collected thoroughly using cutters ($0.10 \times 0.10 \times 0.03$ m^3) with the aid of a sharp knife to prevent disturbance to the structure of the natural soil (see Figure. 6);

Figure.6: Sampling procedure: (**a**) sampling in progress (**b**) collected sample.

- specimens were prepared in the laboratory for testing by trimming oversized samples very carefully to the inside dimensions of the shear box (see Figure. 7);

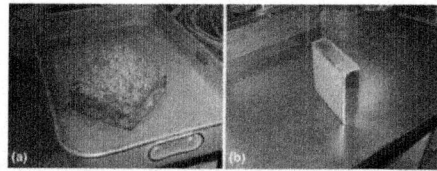

Figure.7: Preparation of specimens for testing: (**a**) sample with protruded material (**b**) specimen after trimming

- after submerging specimens in water, shear tests were conducted at a uniform rate of displacement of 0.03 mm/min, which is assumed to have been slow enough to ensure shear testing under drained conditions.

As a final remark, note that although the standard direct shear box is not the most accurate laboratorial shear apparatus, it is important to underline that its use intends to quantify the natural variability of the material and not necessarily the exact value of the property. So, if all samples are subjected to the same experimental procedure and results are treated equally, then the differences between the obtained parameters are mainly due to natural variability, since methodology errors are systematic.

Natural Variability of Soil Skeleton

During specimen preparation for direct shear tests, two properties which define its physical state in situ were determined—natural unit weight and moisture content. Nevertheless, the evaluation of the natural variability of a soil skeleton, particularly in residual soils, should be carried out using parameters not dependent on the water content, such as dry unit weight and voids ratio. These parameters were determined for each sample and subsequently analysed using statistical procedures (see Table 3, where saturated unit weight is also shown due to its importance in several geotechnical problems).

Table 3: Statistical parameters related to dry and saturated unit weight and voids ratio

Property	N. Samples	$\mu^\wedge\mu^\wedge$	cv^cv ^(%)	Minimum	Maximum
Dry unit weight	42	16.4 kN/m³	3.5	15.1 kN/m³	17.8 kN/m³
Voids ratio	42	0.58	9.6	0.45	0.72
Saturated unit weight	42	20.0 kN/m³	1.8	19.1 kN/m³	20.8 kN/m³

Note that the values for the coefficient of variation of the dry unit weight and the voids ratio are in accordance with the ranges presented in Table 1 for sands and clays. As sedimentary soils are typically more uniform, higher values could be expected in residual soils. However, it is important to acknowledge that unlike clays, where there is a strong correlation between water content and voids ratio, the skeleton of residual soils is significantly more compact and rigid; thus, in residual soils, voids ratio is not so dependent on water content.

In modeling the variables distribution, Kolmogorov–Smirnov Test has proved that normal distribution fits adequately to the set of 42 observations of each three physical properties in question. In fact, the *p-value*, that is, the probability of obtaining a test statistic at least as extreme as the one that has actually been observed under the hypothesis of assuming that variables are, in this case, normally distributed, is higher than the level of significance, , commonly adopted, 5 % (Weber et al. 2006), as it is shown in Table 4.

Table 4: Appropriate statistical laws to model the variability of physical properties

Property	Distribution	*p -value*(%)	α (%)	Kolmogorov–Smirnov Test
Dry unit weight	Normal	53	5	Hypothesis not rejected
Voids ratio	Normal	25	5	Hypothesis not rejected
Saturated unit weight	Normal	84	5	Hypothesis not rejected

Natural Variability of Shear Strength

The characterization of shear strength for the range of normal stresses chosen, 25–100 kPa, was carried out using sets of four adjacent samples (see Figure. 5), assuming the standardization of soil behavior within the area of each set, that is, considering the soil does not hide eventual heterogeneities. For each set, both peak and constant volume shear strength have been estimated by Mohr–Coulomb criterion. The maximum stress of 100 kPa has been adopted since the maximum in situ stress was roughly of this order (see Figure. 8). In such conditions, it was guaranteed that the soil skeleton was not broken during the consolidation phase of the direct shear test.

Figure.8: Excavation slope and location of the study area.

All shear tests were carried under drained conditions to which correspond effective shear strengths; henceforth, to avoid repetitions, the word "effective" is going to be left out. In every single test, the peak and constant volume shear strength were recorded, as well as the normal stress under which the test was conducted. Afterwards, applying the Least-Squares Method to the points which represent the state of stress at failure, two Mohr–Coulomb envelopes were obtained for all sets of samples. In Table 5, relevant statistical data related to the experimental results is shown, namely the parameters defining the Mohr–Coulomb failure envelope for both peak and constant volume strength.

Table 5: Statistical parameters related to peak and constant volume strength

Property	N. Sets of Samples	$\mu^\wedge\mu^\wedge$	$cv^\wedge cv^\wedge (\%)$	Minimum	Maximum
Peak friction angle	10	40.3°	7.9	36.3°	45.6°
Cohesion	10	9.3 kPa	68.0	0.2 kPa	18.9 kPa
Constant volume friction angle	10	37.5°	3.4	35.3°	39.4°

The coefficient of variation for peak friction angle of the granite residual soil from Porto is acceptable according to the benchmarks presented in literature for sandy soils (see Table 1). On the other hand, cohesion presents very random values, with a coefficient of variation close to the upper bound indicated by Shahin and Cheung (2011) for sands and clays—70 %. Consequently, and given that the variability of constant volume shear strength is almost insignificant, it has been concluded that the variability of the fabric of residual soils is what contributes the most for the uncertainties related to their geomechanical characteristics. This variability results also from the nonlinearity of the Mohr–

Coulomb failure criterion, particularly for low normal stresses. In any case, since Mohr–Coulomb failure criterion is so popular for practical purposes, it is important to take note of the high coefficients of variation associated with the use of this model, even having in mind that the values presented for cohesion and peak friction angle should only be considered valid for the range of tested normal stresses.

As to the statistical law that governs the distribution of both peak and constant volume friction angles, the hypothesis of normality has not been rejected by Komogorov-Smirnov Test, as it can be confirmed in Table 6.

Table 6: Suitable statistical laws to model the variability of friction angles

Property	Distribution	p-value (%)	α (%)	Kolmogorov–Smirnov Test
Peak friction angle	Normal	93	5	Hypothesis not rejected
Constant volume friction angle	Normal	88	5	Hypothesis not rejected

However, the set of ten results obtained for cohesion make it difficult to find a statistical law fitting its statistical distribution (see Figure. 9).

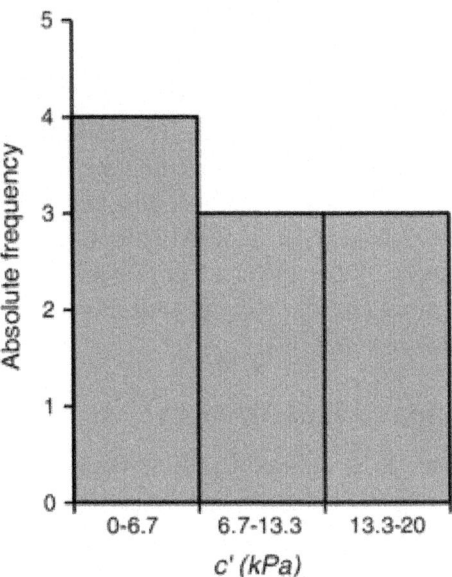

Figure.9: Histogram of the cohesion of the ten sets of samples.

Indeed, even the lognormal distribution suggested by Forrest and Orr (2010) does not fit the results distribution, since, among other reasons, it does not allow the cohesion to assume null values. In order to work around this problem, a hypothetical scenario where only four shear tests had been made to characterize this soil mass was considered. For this hypothetical case, and taking into account that there are 40 experimental results, 10 for each normal stress, it is possible to proceed to $10 \times 10 \times 10 \times 10 = 10000$ random groupings of four samples. The dispersion concerning this scenario is represented in Figure. 10.

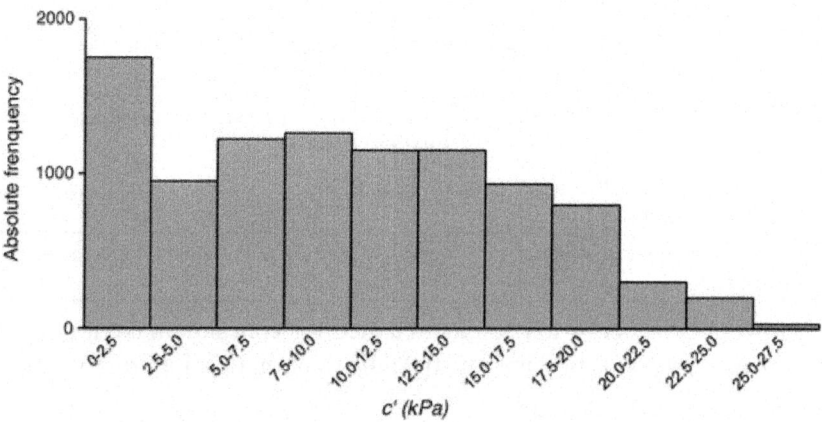

Figure.10: Histogram of cohesion with random groupings of four samples.

The histogram represented in Figure. 10 shows a decreasing tendency of the probability of occurrence as the cohesion increases. Moreover, the class of values lower than 2.5 kPa has the highest absolute frequency, including 916 null cohesions, approximately 10 % of the values obtained by this procedure. This is a consequence of the physical impossibility of negative cohesions and thus, it seems more appropriate and cautious to consider an exponential distribution to model its variability.

Scale of Fluctuation of Peak Strength

When the shear strength is defined by Mohr–Coulomb failure criterion, the peak friction angle and the cohesion cannot be detached, because it depends on the combination of these two parameters. In fact, cohesion and peak friction angle are correlated and, in this particular soil, there is a strong negative correlation defined by a coefficient of correlation, R, of approximately -0.9 (see Figure. 11).

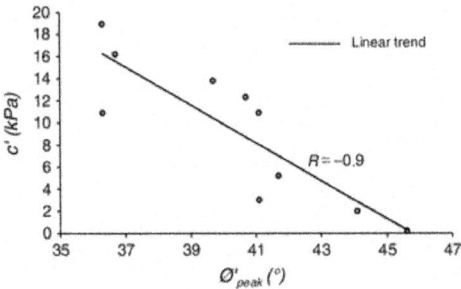

Figure.11: Correlation between peak friction angle and cohesion.

As a result, the comparison between the shear strength exhibited by each set of four samples should be made using another parameter, for example the secant friction angle suggested by Bolton (1986) for sands.

The secant friction angle can be interpreted as a normalization of shear strength, since it is defined as the angle associated with the ratio between the shear strength exhibited by a sample and the normal stress under which the test was carried out. Since shear tests were conducted under the four normal stresses considered for every set of samples, the arithmetic mean of the four secant peak friction angles is representative of the "average" peak shear strength of each sample for the tested range of normal stresses. The map of spatial variability of the secant peak friction angle is shown in Figure. 12.

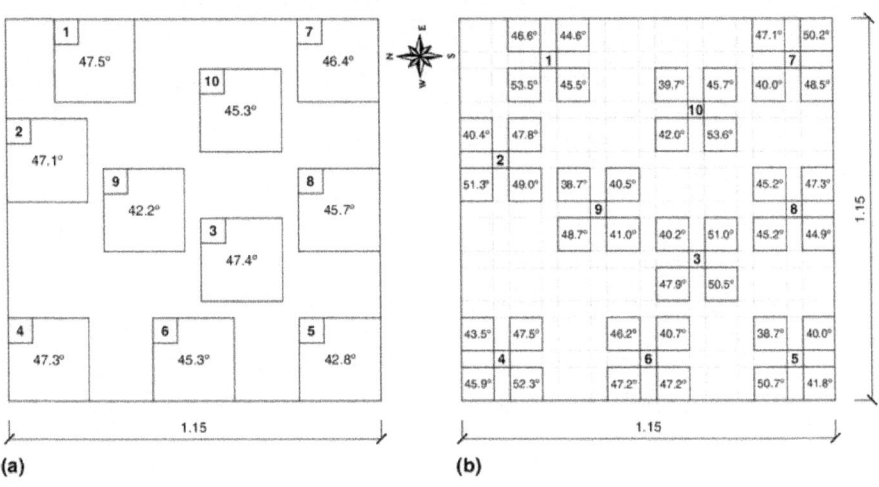

(a)   (b)

Figure.12: Maps of spatial variability of the secant peak friction angle: **(a)** by set of samples (average strength) **(b)** by sample.

As Figure. 12b shows, the mean value of the secant peak friction angle and subsequently the variability of the peak shear strength does not present a dispersion as noteworthy as the one related to the cohesion. Moreover, spatial fluctuations are both random and almost negligible.

So, in order to objectify the spatial component of the natural variability of the peak shear strength, the value of the horizontal scale of fluctuation related to the "average" secant peak friction angle has been determined by the following expedited method.

First, several polylines were defined by an almost random process of grouping sets of samples located within the area in analysis. Those polylines were defined by connecting the geometrical centres of some sets of four samples which characterize areas of 0.25 × 0.25 m². And so, the spatial fluctuations of the "average" peak shear strength within the area of 1.15 × 1.15 m² can somehow be perceived by analyzing the fluctuations of the "average" secant peak friction angle along those polylines.

Furthermore, the quantification of the horizontal scale of fluctuation of the "average" peak shear strength requires the definition of a representative value of the shear strength of the whole area in analysis. As, excluding heterogeneity, the shear strength exhibited by a soil at the same depth and thereafter under the same confining stress should be roughly constant, it seems reasonable to consider the arithmetic mean of the 10 "average" secant peak friction angles as an appropriate reference value for the purpose of estimating the scale of fluctuation.

Hence, considering three different combinations of sets of samples (7-10-9-3-5, 1-2-9-3-5 and 1-10-8-3-6-4, identified in Figure. 12), the horizontal scale of fluctuation of the peak shear strength was estimated using eq. (2)— 0.37, 0.29 and 0.42 m, respectively. As an example, Figure. 13 shows the fluctuations of the "average" secant peak friction angle for the combination of sets of samples resulting in a scale of fluctuation of 0.42 m.

Cautiously, it can be said that an indicative value of the horizontal scale of fluctuation of the peak shear strength of the soil in analysis is 0.4 m, which is notably lower than the ones referred to other geotechnical parameters that characterize the shear strength of both sandy and clayey soils (see Table 2).

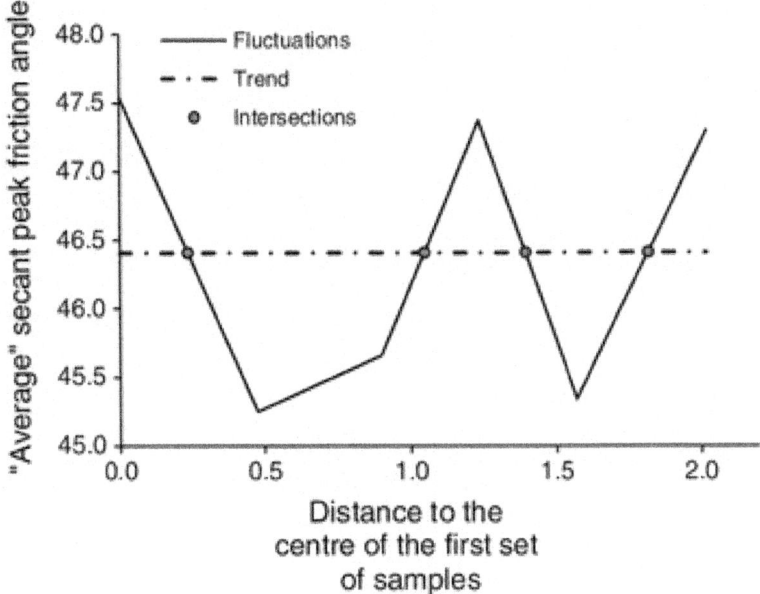

Figure.13: Deviations from trend of the "average" secant peak friction angle along the following sets of samples: 1, 10, 8, 3, 6 and 4.

Scale of Fluctuation of Constant Volume Strength

The same procedure has been applied to the "average" secant constant volume friction angle and the values obtained for the corresponding horizontal scale of fluctuation are 0.40, 0.52 and 0.33 m, respectively, for the following sets of samples: 7-10-9-3-5, 1-2-9-3-5 and 1-10-8-3-6-4. Hence, the horizontal scale of fluctuation of the constant volume shear strength can be taken as 0.5 m.

Influence of Scales of Fluctuation in Coefficients of Variation

As a consequence, the coefficients of variation of both peak and constant volume shear strength parameters, that is, peak and constant volume friction angles and cohesion, should be reduced according to Eq. (3). That is, when analysing the stability of a mass of this particular residual soil along a given failure surface, the coefficients of variation to be considered should be the ones presented in Figure. 14 and Figure. 15, which depend on the length of the failure surface itself. Note that, to simplify this proposal, it is assumed that the vertical scale of fluctuation of the property in analysis is equal to the horizontal one, which seems to be an appropriate assumption for residual soils.

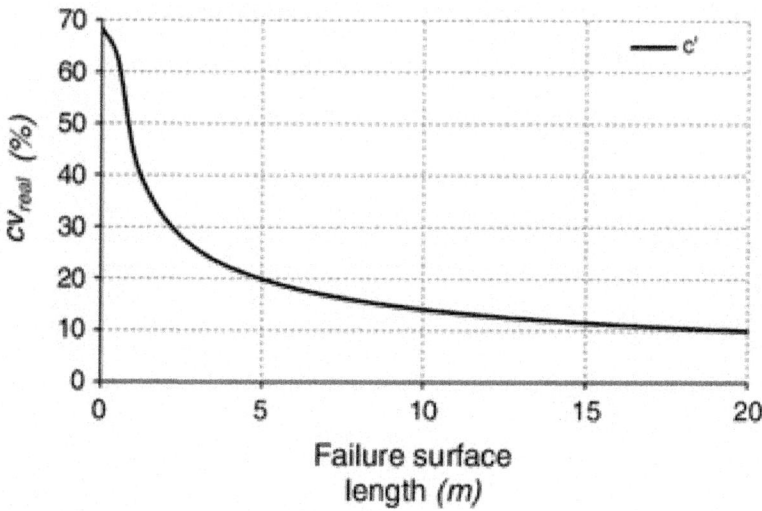

Figure.14: Influence of the scale of fluctuation in the reduction of the coefficient of variation of the cohesion.

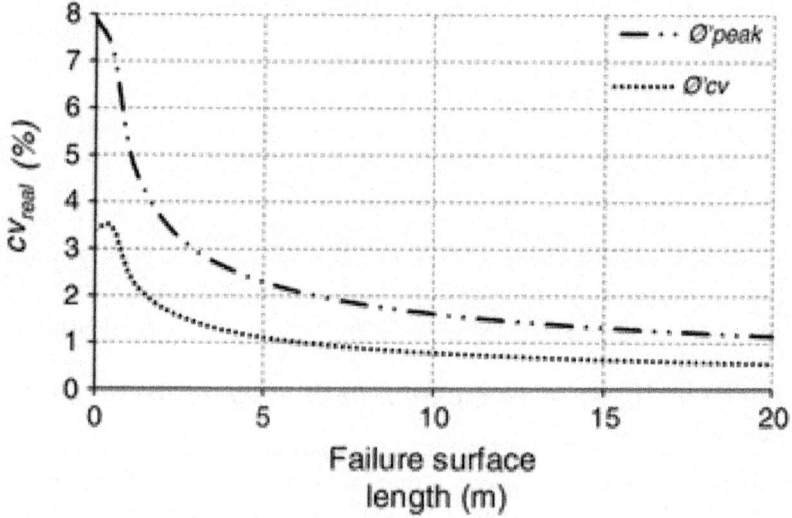

Figure.15: Influence of the scale of fluctuation in the reduction of the coefficients of variation of the peak and constant volume friction angles.

However, it should be noted that the determined scales of fluctuation cannot be higher than the dimensions of the area used to their determination. As this area is particularly small to be considered a representative sample of this highly heterogeneous granitic soil, the estimates presented for both scales of fluctuation do not obviously intend to settle definitive benchmarks for this type of soils. Its purpose is only to contribute for the development of the characterization of the spatial variability of residual soils in general by presenting the analysis of the results of shear tests run on 40 samples of this particular soil.

Characteristic Values

According to clause 2.4.5.2(2) of EN 1997-1:2004/AC:2009, "the characteristic value of a geotechnical parameter shall be selected as a cautious estimate of the value affecting the occurrence of the limit state". Therefore, when the scale of fluctuation of shear strength is small enough in comparison with the length of a given failure surface such that shear resistance is governed by its average value, it is reasonable to consider it as a cautious estimate of the overall shear strength instead of 5 % fractile proposed in clause 4.2(3) of EN 1990:2002/A1:2005/AC:2010. That is, in the particular case of geotechnical structures involving materials with similar geomechanical properties to this granite residual soil from Porto, the characteristic value of shear strength can be considered equal to its mean value, as long as local failures are not a matter of concern.

CONCLUSIONS

The granite residual soil from Porto characterized in this paper presents an important lithological heterogeneity, which is very common in these geomaterials and, in this particular case, easily perceived to the naked eye. The coefficients of variation obtained for both physical and mechanical properties are in accordance with the benchmarks presented in the literature for sedimentary soils, specifically for sands, which represent the main granulometric fraction of residual soils from granite. However, the set of values related to effective cohesion are really scattered, with a coefficient of variation that is very close to the maximum upper limit proposed in the literature for clays. Therefore, it might be more accurate and prudent to use an exponential distribution instead of a lognormal law to model the variability of the effective cohesion of a residual soil, mainly because it allows this parameter to assume null values.

Moreover, as the constant volume shear strength depends mostly on pure friction and its coefficient of variation is slightly lower than the one related to the peak shear strength, it could be concluded that the variability of the shear strength is mainly related to the variability of its fabric, which is destroyed when reaching the peak resistance.

Lastly, it is important to note that the values of the horizontal scales of fluctuation of the peak and the constant volume shear strength of the residual soil in question, approximately and respectively 0.4 and 0.5 m, determined according to the simplified procedure suggested by Kulhawy and Phoon (1999), are extremely small when compared with the benchmarks proposed in the literature for other geomechanical parameters of sedimentary soils (see Table 2). Consequently, the fluctuations around the average shear strength are very persistent along short distances which means that, when failure surfaces are long in comparison with the value of the scale of fluctuation, the shear strength of the ground is controlled by its average value, that is, even if the distribution of the shear strength along the failure surface is highly scattered, redistribution of shear stresses can take place and so compensate the lack of shear strength of the most fragile areas. However, as geotechnical properties of residual soils depend on many factors, their statistics may differ significantly from site to site which means that theses scales of fluctuation might not apply to every single granitic soil. Therefore, further investigations should be carried out in the future in order to complement the main conclusions presented in this paper regarding this issue.

ACKNOWLEDGMENTS

The authors would like to acknowledge the Geotechnical Laboratory of the Engineering Faculty of the University of Porto (LABGEO-FEUP), in the person of Professor Viana da Fonseca, for providing all the required resources to carry out the experimental tests. Grateful acknowledgements extend to Geologist Lígia Santos for carrying out the geological cartography of the soil analysed in this paper.

REFERENCES

1.   ASTM D3080-04. Standard test method for direct shear test of soils under consolidated drained conditions. ASTM International

2.   Baecher G, Christian J (2003) Reliability and statistics in geotechnical engineering. John Wiley & Sons Ltd., Chichester, England

3.   Bolton M (1986) The strength and dilatancy of sands. Géotechnique 36(1):65–78

4.   Curto J, Pinto J (2009) The coefficient of variation asymptotic distribution in the case of non-iid random variables. J Appl Stat 36(1):21–32

5.   Duncan J (2000) Factors of safety and reliability in geotechnical engineering. J Geotech and Geoenvironmental Eng 126(4):307–316

6.   EN 1990:2002/A1:2005/AC:2010. Eurocode 0: Basis of structural design. CEN

7. EN 1997-1:2004/AC:2009. Eurocode 7: Geotechnical design—part 1—general rules. CEN

8. Forrest W, Orr T (2010) Reliability of shallow foundations designed to Eurocode 7. Georisk 4(4):186–207

9. Google Maps (https://www.google.com/maps). Accessed on 2014-04-22

10. Kulhawy F, Phoon K–K (1999) Characterization of geotechnical variability. Can Geotech J 36(4):612–624

11. Pinheiro Branco L (2011) Aplicação de conceitos de fiabilidade a solos residuais. MSc thesis, Faculty of Engineering, University of Porto, Portugal (in Portuguese)

12. Shahin M, Cheung E (2011) Stochastic design charts for bearing capacity of strip footings. Geomech Eng 3(2):153–167

13. Silva Cardoso A, Matos Fernandes M (2001) Characteristic values of ground parameters and probability of failure in design according to Eurocode 7. Géotechnique 51(6):519–531

14. Vanmarcke E (1977) Probabilistic modeling of soil profiles. J Geotech Eng Division, ASCE 103(GT11):1227–1246

15. Viana da Fonseca A, Matos Fernandes M, Silva Cardoso A (1997) Interpretation of a footing load test on a saprolitic soil from granite. Géotechnique 47(3):633–651

16. Viana da Fonseca A, Rios Silva S, Cruz N (2010) Geotechnical characterization by in situ and lab tests to the back-analysis of a supported excavation in metro do porto. Geotech Geol Eng 28(3):251–264

17. Weber M, Leemis L, Kincaid R (2006) Minimum Kolmogorov-Smirnov test statistic parameter estimates. J Stat Comput Simul 76(3):195–206

Chapter 8

ELECTROKINETIC STABILIZATION OF SOFT SOIL USING CARBONATE-PRODUCING BACTERIA

Hamed A. Keykha[1], Bujang B. K. Huat[1], Afshin Asadi[2]

[1]Huat Department of Civil Engineering, Faculty of Engineering, Universiti Putra Malaysia (UPM), 43400 Serdang, Selangor, Malaysia

[2]A. Asadi (&) Faculty of Engineering, Housing Research Center, Universiti Putra Malaysia (UPM), 43400 Serdang, Selangor, Malaysia

ABSTRACT

The carbonate (CO_3^{-2}) produced by *Sporosarcina pasteurii* was injected electrokinetically to enhance the mechanical properties of soft clay soils. In this method the Ca^{2+} was injected into the anode chamber and moved towards the cathode by electromigration and electroosmotic flow. Then the released CO_3^{-2} from a blend of bacteria and urea was injected into the cathode chamber. The CO_3^{-2} ions were moved from the cathode to the anode under electromigration mechanism. The $CaCO_3$ was precipitated in the presence of calcium in porous medium of the soil, and consequently increased the shear strength of the soil. The polarity reversal was applied to have a homogeneous distribution of $CaCO_3$.

INTRODUCTION

Electrokinetic (EK) stabilization is a ground improvement technique for treating subsurface soils without excavation, unlike conventional methods. Several researchers have studied the applications of EKs principles on engineering properties of soils and soil improvement (Esrig and Gemeinhardt 1967; Johnston and Butterfield 1977; Shang and Dunlap 1996; Ozkan et al. 1999; Barker et al. 2004; Ou et al. 2009; Chien et al.2012). The injection of agents under electric fields could be used to overcome problems in heterogeneous and/ or low permeability soils (Alshawabkeh 2001). The movement of stabilizing agents into the soil mass is governed by the principles of EKs, whereas the mechanisms of stabilization can be explained by the principles of chemical stabilization. When ions are used as stabilizing agents, the ions migrate into soils through processes of electro migration and osmotic advection. These ions improve the soil strength by three mechanisms, ion replacement, mineralisation and precipitation of species in the pore fluid. It is precipitation that provides to the greatest contribution to increase of the strength (Mohamedelhassan and Shang 2002; Alshawabkeh and Sheahan 2003; Asavadorndeja and Glawe 2005; Barker et al. 2004).

Biogrouting is a biological ground improvement method, focuses on microbially induced carbonate precipitation (MICP), in which microorganisms are used to induce carbonate precipitation in the subsurface in order to increase the strength and stiffness of granular soils (Whiffin 2004; De Jong et al. 2006, 2010; Whiffin et al. 2007; Ivanov and Chu 2008; Van Paassen et al. 2010).

Most studies on biogrouting use microorganisms containing the enzyme urease, in particular, the bacterium of *Sporosarcina pasteurii*. The microbial urease catalysis the hydrolysis of urea into ammonium and carbonate ($CO_3{}^{2-}$) (Eq. 1).

$$CO(NH_2)_2 + 2H_2O \rightarrow 2NH_4^+ + CO_3^{-2}$$
(1)

The produced carbonate ions precipitate in the presence of calcium ions as calcium carbonate crystals ($CaCO_3$), which form cementing bonds between the grains of the soil (Eq. 2).

$$Ca^{2+} + CO_3^{-2} \rightarrow CaCO_3 \downarrow$$
(2)

There are some limitations related to the bacterial injection in fine grain soils when it is pumped. As bacteria have a typical size of 0.5–3 µm, it is hard to be transported through silty or clayey soils. Bioclogging could occur when bacteria are adsorbed or strained by the solid grains, resulting a heterogeneous $CaCO_3$ precipitation in the soil (Mitchell and Santamarina 2005).

Previous research of biogrouting investigated the improvement of coarse grain soils (Whiffin 2004; De Jong et al. 2006; Whiffin et al. 2007; Ivanov and Chu 2008; Harkes et al. 2010; Van Paassen et al. 2009, 2010). As the permeability of coarse soil is high, passing the bacteria through the soil is possible. However, in fine grain soil, because the permeability of the soil is low, the transport of bacteria is difficult. Subsequently, the precipitation of $CaCO_3$ in porous media in the biomineralisation process could be heterogeneous. Furthermore, the control of grout pressure and pumping rate is difficult in various types of soils (i.e., fine and coarse soils) and depth. When high-pressure pumps are used to inject fluids into the formation, as pressure builds, soil structure fails and the fluid causes soil to separate or "fracture". As the injection progresses, this separation propagates through the formation in a very unpredictable manner. Moreover, the flow velocity of grout can effect $CaCO_3$ distribution. The lack of $CaCO_3$ close to the injection point in the experiments could be the result of higher flow velocity, causing more bacterial flush out and hence, lower activity and less $CaCO_3$ precipitation. Another explanation for the lack of $CaCO_3$ around the injection points, considers the kinetics of $CaCO_3$ precipitation and the transport of crystals (Van Paassen et al. 2010).

The aim of this study is EK stabilization of soft clay soil using the product of bacteria (CO_3^{-2}) and calcium (Ca^{2+}).

MATERIALS AND METHODS

Bacterial Growth Condition

The strain *S. pasteurii* (ATCC 11859) was used in this study. The medium used for harvesting *S. pasteurii* was NH4-YE prepared as follows. Yeast extract (20.0 g) was dissolved in 100 ml of distilled water (ingredient A), $(NH_4)_2SO_4$ (10.0 g) in 100 ml of distilled water (ingredient B), and Tris buffer (15.75 g) in 800 ml distilled water (ingredient C). Ingredients A, B and C were then autoclaved separately (121 °C for 15 min). After sterilization, each ingredient was allowed to cool and then the ingredients were mixed in a 1,000 ml Erlenmeyer flask. The NH4-YE medium has a pH value of 9.0, which is appropriate for *S. pasteurii* growth. After, the bacteria were moved to NH4-YE broth in an incubator (30 °C) and shaken at 200 rpm while mouth of the Erlenmeyer flask was plugged with a silicon cork to allow the cultivation under aerobic conditions. The color of the culture broth turned muddy within 2–3 days. An overnight culture was harvested by centrifugation (Sartorius AG, Sigma 3–18 K, Germany) at $8,000g$ for 10 min with an optical density of 0.5 at 600 nm (Labomed UVD 2950) (Mortensen et al. 2011).

Experimental Setup

Figureure 1 shows the EK setup which was made of Plexiglas with 30-cm long, 10-cm wide and 15-cm high The setup was designed with two chambers (6 cm length) in the left and right of the soil specimen holder (18 cm length). Two graphite electrodes (Graphite Laminate SLS) were placed into the chambers. Two Mariote bottles and two pH controllers were connected to each compartment. They maintained a constant water level across the specimen and prevented any hydraulic gradient. The pH controllers with two acid and base tanks were designed to adjust the pH in the chamber solution. 0.5 M NaOH and 0.5 M HCl solutions were used for pH adjustment.

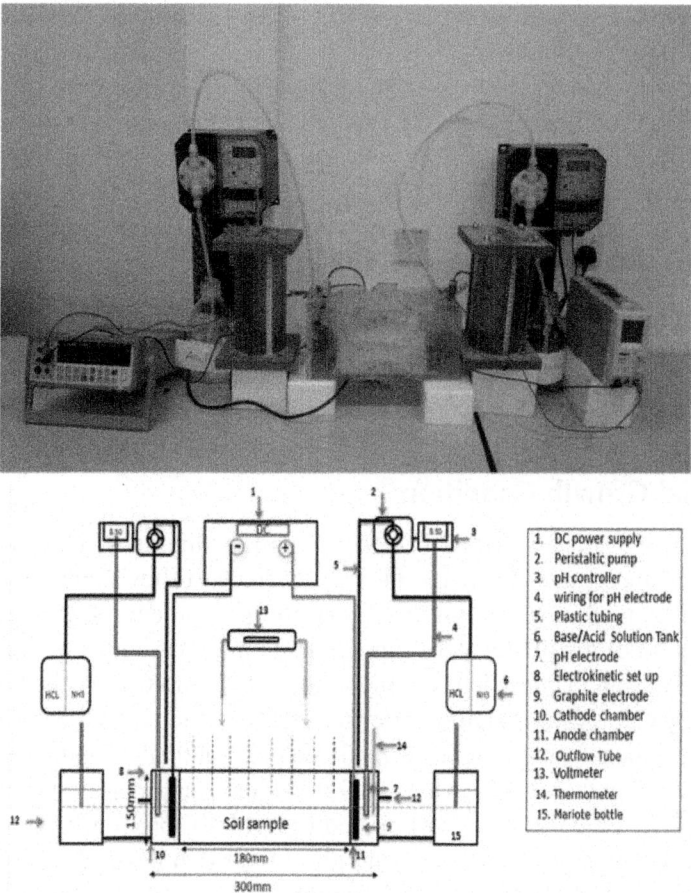

Figure. 1: Experimental setup for EK treatment

Any excess fluid transported owing to electroosmosis flow was collected in cylinders placed at the end of the chambers. Normally, the electroosmotic flow will be from the anode to the cathode. The DC current was applied by a power supply in this study. A voltmeter and a thermometer respectively measured the voltage potential between two points across the specimen and the temperature because of the electrical current in the compartments.

Soil Properties

Table 1 shows the engineering properties of the soil used in this study. This soil had a low hydraulic permeability about 2.7×10^{-7} cm/s. Figureures 2 and 3 illustrate scanning electron microscopy image and energy-dispersive X-ray analysis of kaolinite minerals respectively.

Table 1: The engineering properties of the soil

Properties	Values	Method
Specific gravity	2.6	ASTM D-854
Liquid limit (%)	54	ASTM D-4318
Plastic limit (%)	35	ASTM D-4318
Plasticity index (%)	19	–
Hydraulic conductivity (cm/s)	2.7×10^{-7}	Head (1982)
Maximum dry density (g/cm^3)	1.49	ASTM D-698
Optimum moisture content (%)	31.15	ASTM D-698
Unified soil classification	CH	ASTM D-2487
Mineral Composition	Kaolinite $(Al_2Si_2O_5(OH)_4)$	X-ray analysis
Undrained shear strength (kPa)	6	BS 1377-7

Figure. 2: SEM image of kaolinite minerals.

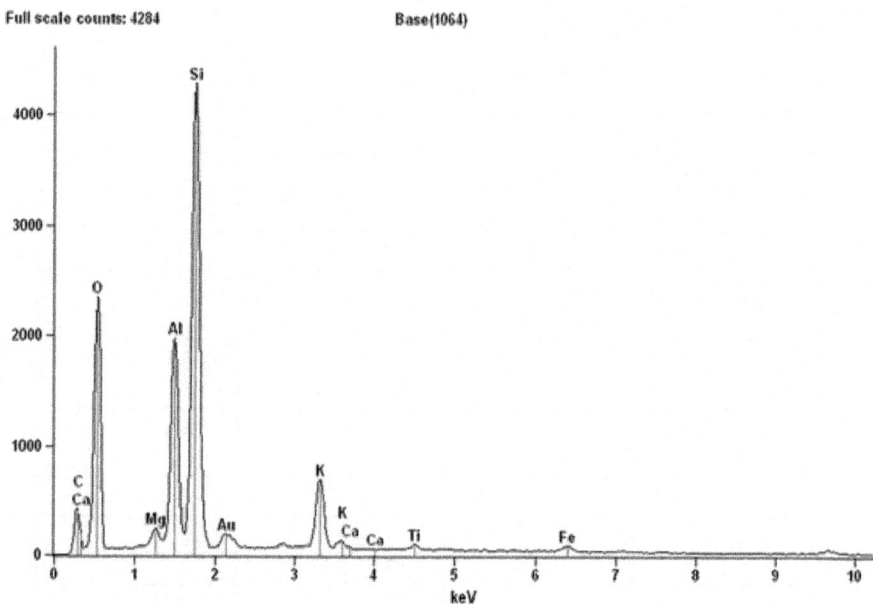

Figure. 3: EDX analysis of the clay soil sample.

EXPERIMENTAL PROCEDURES

Sample Preparation

The soil specimens were prepared at 85 % maximum dry density (1.3 g/cm³) and 21 % of moisture to make the soil more permeable. The soil was placed in the test container as the basis of sample preparation (ASTM D698-07). Filter papers were used at both ends of the specimen to avoid material loss and to prevent contamination of the electrode fluids by soil particles. The soil sample was allowed to saturate by adding distilled water for 48 h before EK treatment.

Electrokinetic Experiments

Table 2 lists the detailed test procedure of the EK treatment. In the experiment, a calcium chloride solution (2 M) was injected into the anode chamber over a period of 3 days (T1). A blend of bacterial suspension and urea (1 M) about 150 ml was prepared and then bacteria were allowed to be activated aerobically. When the bacteria were exposed to the urea, they released the urease enzyme and made carbonate ions (CO_3^{-2}). Then the solution was passed through a filter paper (0.2 μm) allowing the bacteria and particles retained and the carbonate solution remained. The product of bacterial activity (i.e. CO_3^{-2}

solution) was injected into the cathode chamber in treatment time of 4 days for normal polarity and 4 days for reversing polarity in the process (T2 and T3).

Table 2: Summary of test procedure

No	Code	Test activities	Date (days)	Normal polarity	Polarity reversal
1	T1	Injection of the calcium chloride solution into the anode chamber	1–3	√	
2	T2	Injection of the culture bacteria and urea into the cathode chamber	4–7	√	
3	T3	Injection of the culture bacteria and urea into the cathode chamber	8–11		√
4	T4	Measurement of electrical potentials across the specimen and over the time	1–11	√	√
5	T5	Vane shear test across the specimen	18	–	–
6	T6	Measurement of water content and $CaCO_3$ percent on each section	18	–	–

The voltage gradient of 60 mV was held constant in this EK experiment between the anode and cathode (18 cm). During the experiments, electrical potentials over time and across the specimen were monitored (T4). After 7 days curing time, the shear strengths were measured using a 12.7 mm vane across the soil specimen at the horizontal distance of 4, 8, 12 and 16 cm from the cathode to the anode (BS1377-7-3: 1990). For each horizontal distance, the vane was positioned at two points at the same level in order to take the average value of the vane shear strength. The center-to-center distance between the two points of measurements was 4 cm (T5). Water content and percentage of $CaCO_3$ were also measured on each section (T6). The water content of each sample was measured by weight as the ratio of the mass of water present to the oven dry weight of the soil sample. The $CaCO_3$ percentage was determined using the acid washing technique (Mortensen et al. 2011). In this technique, the oven dried mass of the soil samples across the specimen were measured before and after an acid wash in 5 M solution of HCl. The dissolved calcium chloride after treatment was filtered. The difference in the two measured masses before and after treatment was taken as the mass of $CaCO_3$.

RESULTS AND DISCUSSION

Undrained Shear Strength

Undrained shear strength tests demonstrated the improvement of strength and stiffness of the soft clay after EK treatment. The undrained shear strength of the untreated soil was 6 kPa with moisture content 60 % from the cathode to the anode. To examine the effect of EK treatments, the shear resistance of the soil at the horizontal distance of 4, 8, 12 and 16 cm from the cathode was obtained. Figureure 4 shows the undrained shear strength and the corresponding moisture content of the soil which had an increase in strength 10 times after a 7-day curing. The highest improvement in strength was found near the cathode because of the high concentration of carbonate ions (CO_3^{-2}). The water content increased from the anode to the cathode (73–77 %), indicating that the increase in strength is as a result of some chemical reactions.

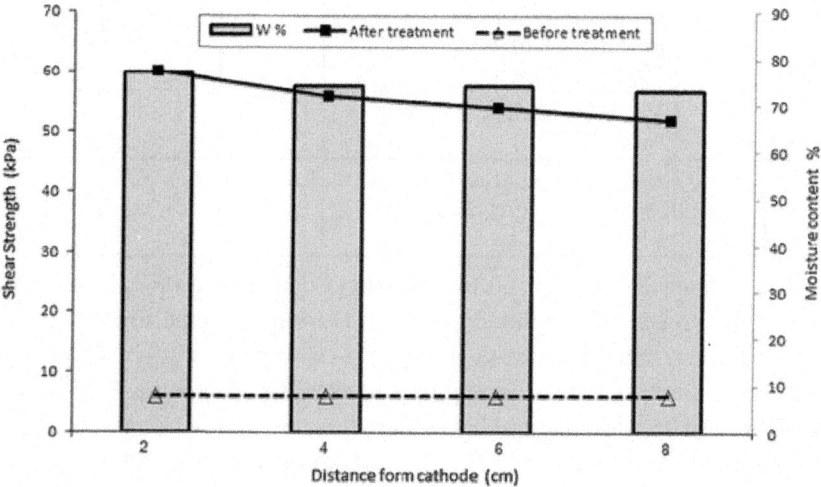

Figure. 4: The undrained shear strength before and after treatment with moisture content

Detection of Calcium Carbonate

Figureure 5 shows the percentage of precipitation $CaCO_3$ across the specimen, which is an evidence for soil improvement during EK treatment. The $CaCO_3$ content was from 10 to 16 % across the soil sample (T6). The increase in shear strength was due to the increase of $CaCO_3$ across the sample. Scanning electron microscopy shows microstructure of calcium carbonate crystals between the soil particles (Figure. 6).

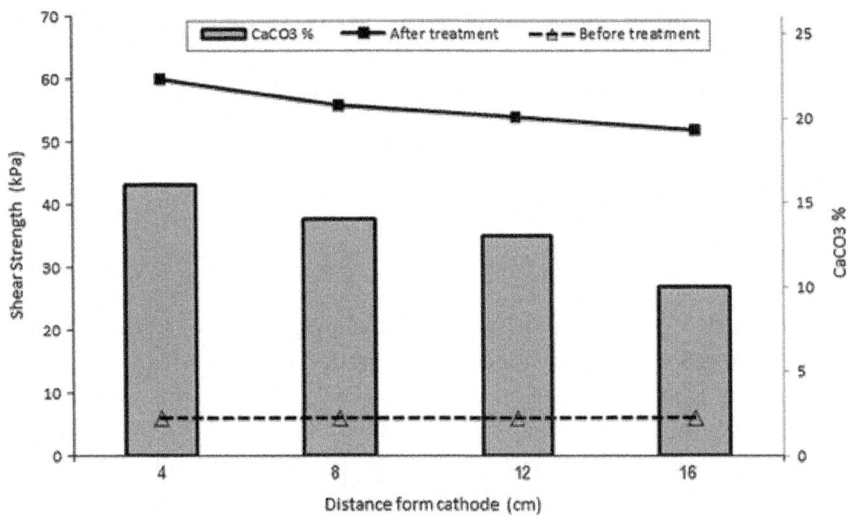

Figure. 5: The percentage of $CaCO_3$ deposits across the specimen relate to shear strength.

Figure. 6: SEM photo of $CaCO_3$ crystals.

Variation of pH in EK Treatment

Figureure 7 shows the pH variation in the compartments during the tests T1, T2 and T3 (i.e. 11 days). The electrolysis of water at the anode and the cathode due to the application of direct electric current through electrodes produces oxygen and hydrogen, respectively (Eqs. 3, 4). Oxidation of water at the anode generates an acid medium, whereas reduction at the cathode produces a basic medium that causes a pH gradient between the two electrodes (Acar et al. 1990).

$$2H_2O - 4e^- \rightarrow O_2 + 4H^+ \text{ (anode)} \tag{3}$$

$$4H_2O + 4e^- \rightarrow 2H_2 + 4OH^- \text{ (cathode)} \tag{4}$$

As the $CaCO_3$ precipitation needs an alkaline condition, two-pH controllers adjusted the pH in both the anode and cathode compartments. The pH value in the pump controller was set at 8.5 for both compartments. There was no need for pH adjustment during T1 experiment (1–3 days), because during this period, Ca^{2+} were injected into the anode chamber. A significant change in pH was also observed in T1, reaching a maximum of 12.5 and a minimum of 1.5 in the cathode and the anode chambers respectively. However, the pH was adjusted during T2 experiment (4–5 days) and in the polarity reversal T3 experiment (6–7 days).

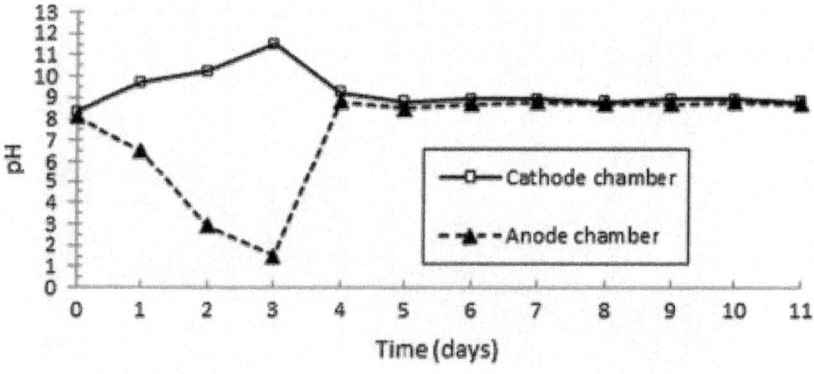

Figure. 7: The pH variation in the cathode and the anode compartments during 11 days of the EK treatment.

Variation of Voltage and Temperature

Figureure 8 shows the variation of voltage distribution with time during the EK treatment. At first day, the voltage distribution was linear between the electrodes in the soil because of the uniform electric conductivity. After 2 days, the voltage increased near the anode due to the presence of calcium ions. The maximum voltage distribution continued due to the electrolysis reactions and the electromigration of Ca^{2+} from the anode to the cathode until 4 days (T1 experiment). During the injection of production of bacteria (T2 experiment), as CO_3^{2-} were induced from the cathode to the anode, an increase of electrical potential was observed initially in the soil (7 days). Then, a zone of high electrical resistance developed near to the cathode. This is probably due to the precipitation of $CaCO_3$ into the soil. During polarity reversal (T3 experiment), the decrease of electrical potential near the cathode demonstrated a higher electrical resistance. After 11 days, a sharp drop in the electrical potential occurred due to the precipitation of $CaCO_3$ across the soil specimen.

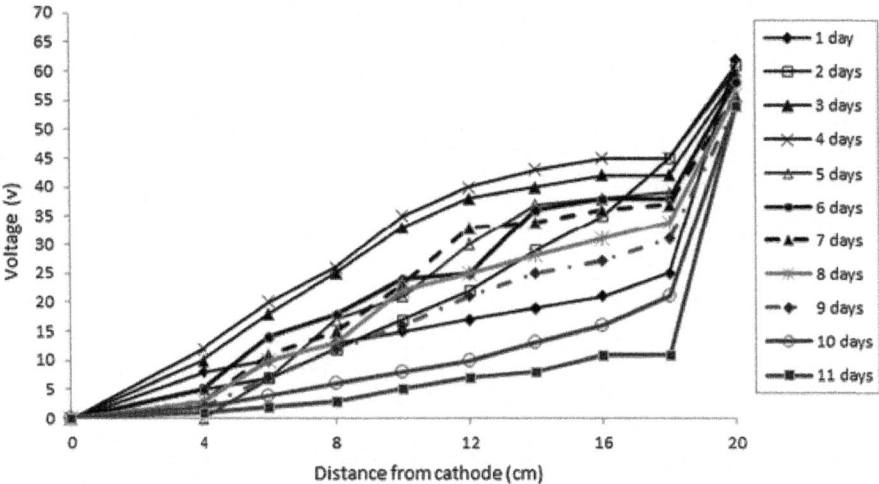

Figure. 8: The voltage variation across the soil sample during 11 days of EK treatment.

EK Treatment Mechanism

The applied electrical potential into the soil was initiated by both electroosmosis, and electromigration. The calcium ions (Ca^{2+}) were moved from the anode chamber through the soil specimen towards the cathode chamber by both electromigration and electroosmosis during 3 days. Figureure 9 illustrates the cumulative volume of the discharge fluid and its cation compositions against elapsed time by the electroosmotic flow in the cathode chamber during 3 days. The carbonate ions that have a negative charge were moved through the soil specimen towards the anode chamber. The carbonate quickly connected with the calcium ions (Ca^{2+}). Ion migration can be used for injection of bacterial product (i.e. CO_3^{-2}) and Ca^{2+} to enhance the stabilization of the soils.

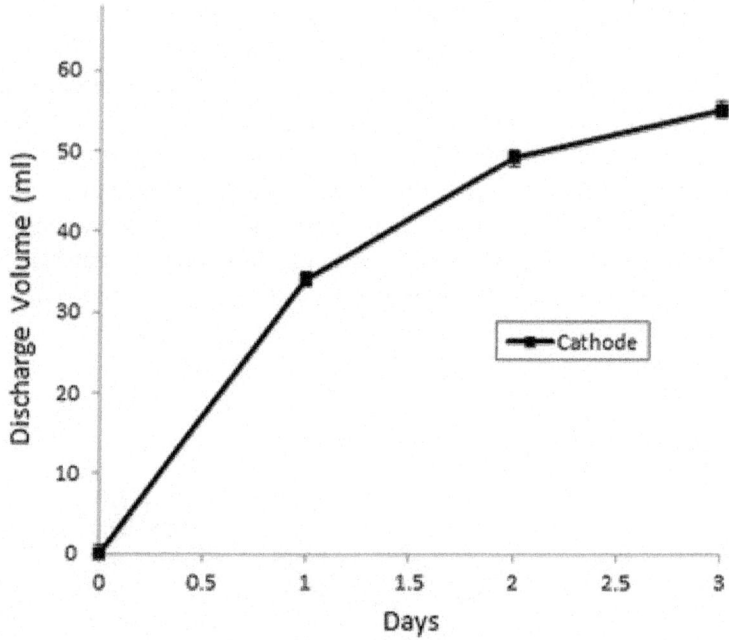

Figure. 9: The discharged volume of electroosmotic flow in the cathode chamber during 3 days.

Previous studies showed that the EKs has delivered significant benefits in terms of both efficiency and cost when applied to difficult ground engineering (Alshawabkeh and Sheahan 2003; Barker et al. 2004; Ahmad et al. 2010; Chien et al. 2012). It can be designed to target a specific area. The energy demands are really relatively small. Without having to excavate the entire area, money and time are saved. Therefore, EK stabilization of soft clay using the production of bacteria is considered a clean method. This method has two advantages. Firstly, the microbial carbonate precipitation is safe and environmental friendly compared to the chemical materials used in EK stabilization such as sodium silicate, aluminum hydrates, phosphoric acid (Asavadorndeja and Glawe 2005; Kamarudin et al. 2006; Ou et al. 2009; Abdullah and Al-Abadi 2010; Chien et al. 2012). Secondly, using bacterial production (i.e. CO_3^{-2}) for deposition of $CaCO_3$ homogeneously in fine soils is more effective compared to biocementation techniques (soil mixing and biogrouting).

CONCLUSIONS

Electrokinetic treatments were conducted on soft clay (kaolinite) using produced carbonate of bacterial activity and calcium. The calcium (Ca^{2+})

electrokinetically was transported by electrooomotic flow and electromigration across the soil specimen from the anode to the cathode. A blend of bacteria and urea solution that released high concentrations of carbonate were filtered and injected into the cathode chamber. As the carbonate ions (CO_3^{-2}) were moved from the cathode to the anode, the $CaCO_3$ was precipitated in the presence of calcium (Ca^{2+}) in the soil. The injection of carbonate solution and calcium could be effectively flowed in soft clay during EK treatment. The water content increased from the anode to the cathode by the electroosmotic flow from 73 to 77 %. Therefore, a significant strength increase up to 60 kPa of untreated soil (10 times) was observed in the soft clay across the specimen, with increasing $CaCO_3$ precipitation. The $CaCO_3$ content was from 10 to 16 % across the soil sample. It can be concluded that the EK stabilization technique is a sustainable and environmental friendly method to improve the mechanical properties of the soft soils due to the microbial carbonate precipitation.

ACKNOWLEDGMENTS

Financial supports from the Research Management Center (RMC) of the Universiti Putra Malaysia under Grant No. 5527094 is gratefully acknowledged.

REFERENCES

1. Abdullah WS, Al-Abadi AM (2010) Cationic–electrokinetic improvement of an expansive soil. Appl Clay Sci 47(3–4):343–350

2. Acar YB, Gale RJ, Hamed J, Putnam G (1990) Electrochemical processing of soils; theory of pH gradient development by diffusion and linear convection. J Environ Sci Health A 25(6):687–714

3. Ahmad KB, Taha MR, Kassim KA (2010) Electrokinetic treatment on a tropical residual soil. Proc ICE-Ground Improv 164(1):3–13

4. Alshawabkeh AN (2001) Basics and applications of electrokinetic remediation. Handouts prepared for short course. Federal University of Rio de Janeiro, Rio de Janeiro, Brazil, 19–20 Nov 2001

5. Alshawabkeh AN, Sheahan TC (2003) Soft soil stabilization by ionic injection under electric fields. Ground Improv 7:177–185

6. Asavadorndeja P, Glawe U (2005) Electrokinetic strengthening of soft clay using the anode depolarization method. Bull Eng Geol Environ 64:237–245

7. ASTM (2007) Moisture–density relations for soils and soil–aggregate mixtures. D698-07e1, American Society for Testing and Materials 04.08, USA

8. Barker JE, Rogers CDF, Boardman DI, Peterson J (2004) Electrokinetic stabilization: an overview and case study. Ground Improv 8(2):47–58

9. British Standard Institution (1990) Methods of test for soils for civil engineering purposes. BSI1377, part 1–9, HMSO, London

10. Chien SC, Ou CY, Lo WW (2012) Electro-osmotic chemical treatment of clay with interbedded sand. J Geotech Eng ASCE 167(1):1–10

11. De Jong JT, Fritzges MB, Nusslein K (2006) Microbially induced cementation to control sand response to undrained shear. J Geotech Geoenviron Eng 132(11):1381–1392

12. De Jong JT, Mortensen BM, Martinez BC, Nelson DC (2010) Bio-mediated soil improvement. Ecol Eng 36(2):197–210

13. Esrig MI, Gemeinhardt JP Jr (1967) Electrokinetic stabilization of an illitic clay. J Soil Mech Found Div ASCE 93(SM3):109–128

14. Harkes MP, Van Paassen LA, Booster JL, Whiffin VS, Van Loosdrecht MCM (2010) Fixation and distribution of bacterial activity in sand to induce carbonate precipitation for ground reinforcement. Ecol Eng 36(2):112–117

15. Head KH (1982) Manual of soil laboratory testing. Pentech Press, London

16. Ivanov V, Chu J (2008) Applications of microorganisms to geotechnical engineering for bioclogging and biocementation of soil in situ. Rev Environ Sci Biotechnol 7(2):139–153

17. Johnston IW, Butterfield R (1977) A laboratory investigation of soil consolidation by electro-osmosis. Aust Geomech J G7:21–32

18. Kamarudin A, Anuar KK, Raihan TM (2006) Electroosmotic flows and electromigrations during electrokinetic processing of tropical residual soil. Malays J Civil Eng 18(2):74–88

19. Mitchell JK, Santamarina JC (2005) Biological considerations in geotechnical engineering. J Geotech Geoenviron Eng 131(10):1222–1233

20. Mohamedelhassan E, Shang JQ (2002) Effect of electrode materials and current intermittence in electro-osmosis. Ground Improv 5:3–11

21. Mortensen BM, Haber MJ, DeJong JT, Caslake LF, Nelson DC (2011) Effects of environmental factors on microbial induced calcium carbonate precipitation. J Appl Microbiol 111(2):338–349

22. Ou C-Y, Chien S-C, Chang H-H (2009) Soil improvement using electroosmosis with the injection of chemical solutions: field tests. Can Geotech J 46(6):727–733

23. Ozkan S, Gale RJ, Seals RK (1999) Electrokinetic stabilization of kaolinite by injection of Al and PO$_4$$^{-3}$ ions. Ground Improv 3(4):135–144

24. Shang JQ, Dunlap WA (1996) Improvement of soft clays by high voltage electrokinetics. J Geotech Eng ASCE 122(4):274–280

25. Van Paassen LA, Harkes M P, Van Zwieten GA, Van der Zon WH, Van der Star WRL, Van Loosdrecht MCM (2009) Scale up of BioGrout: a biological ground reinforcement method. In: Hamza M, Shahien M, Mossallamy YE (eds) Proceedings of 17th international conference on soil mechanics and geotechnical engineering (ICSMGE), pp 2328–2333

26. Van Paassen LA, Daza CM, Staal M, Sorokin DY, Van der Zon W, Van Loosdrecht MCM (2010) Potential soil reinforcement by microbial denitrification. Ecol Eng 36(2):168–175

27. Whiffin VS (2004) Microbial CaCO3 precipitation for the production of biocement. Ph.D. thesis, School of Biological Sciences and Biotechnology, Murdoch University, Perth Australia

28. Whiffin VS, Van Paassen LA, Harkes MP (2007) Microbial carbonate precipitation as a soil improvement technique. Geomicrobiol J 24(5):417–423

Chapter 9

TREATMENT OF COLLAPSIBILITY OF GYPSEOUS SOILS BY DYNAMIC COMPACTION

Mohammed Y. Fattah[1], Hawraa H. M. al-Musawi [2], Firas A. Salman[3]

[1]Building and Construction Engineering Department, University of Technology, Baghdad, Iraq

[2]Civil Engineering Department, College of Engineering, University of Kufa, Kufa, Iraq

[3]Department of Civil Engineering, Faculty of Engineering, University of Malaya, 50603 Kuala Lumpur, Malaysia

ABSTRACT

Gypseous soils are distributed in vast areas and various regions of Iraq and other countries. Many foundation failure problems that occur in these soils are associated with percolation of water and dissolution of gypsum. Many attempts were made by several researchers to treat and improve the properties of gypseous soils to decrease the dissolution of gypsum and collapse potential of these soils. The purpose of the present work is to investigate the effect of dynamic compaction process on the behaviour of gypseous soils. Extensive laboratory tests are carried out to study the geotechnical properties and the behaviour of three gypseous soils of different gypsum contents; 60.5, 41.1 and 27 %. The tests included compaction characteristics, compressibility, and collapsibility tests for samples tested before and after treatment by dynamic compaction process under different number of blows, falling weights and heights of falling of the weights. Three weights are used to compact the samples, namely; 2, 3 and 5 kg. The number of blows is varied between 20 and 40, while three heights of drop are tried (35, 50 and 65) cm. The results showed that the best improvement in compressibility is achieved when the sample is compacted by 20 blows; above this number a negligible decrease in the compression index C_c is obtained. As the gypsum content increases, the dynamic compaction has greater effect on improvement of compressibility of the soil, while as the height of drop increases, the compression index C_C decreases.

INTRODUCTION

Gypsiferous soils are soils that contain sufficient quantities of gypsum (calcium sulphate) to interfere with plant growth (Nettleton et al. 1982). They occupy about 90 million ha across Algeria, Argentina, Australia, Iraq, Libya, Somalia, Spain, Sudan, Syria, the former USSR and other arid and semi-arid countries with annual rainfall of less than 500 mm (FAO 1990).

In Iraq, gypseous soils cover wide areas, sometimes with high gypsum content that exceeds the soil content, and today engineering properties of these soils in some areas are unknown. The basin of Iraq covers more than about 30 % of Iraqi area (Nashat 1994).

Gypseous soils usually stiff when they are dry, but these soils may be affected greatly when subjected to changes in water content due to water table fluctuation, or due to water infiltration which may dissolve gypsum causing, pores, cracks and producing cavities that lead to increase the permeability in gypseous soils. Therefore, the safety and good performance of the foundations of structures and earth structures, such as embankments and dams, will be governed by the changes in the properties of these soils.

Rollins et al. (1998) evaluated the influence of moisture content on dynamic compaction efficiency at six field test cells, each with a progressively higher average moisture content. The soil profile consisted of collapsible sandy silt, and average test cell moisture contents ranged from 6 to 20 %. At each cell, compaction was performed with a 4.54 ton weight dropped from a height of 24.3 m. Compaction efficiency was evaluated using (1) crater depth measurements, (2) cone penetration tests before and after compaction, and (3) undisturbed samples before and after compaction. It was found that crater depth increased by a factor of 4 as moisture content increased. The degree of improvement increased up to a moisture content of about 17 % and then decreased. The optimum moisture content and the maximum dry unit weight are similar to those predicted by laboratory Proctor testing using energy levels comparable to those employed in the field. Maximum dry unit weight decreased with depth, while optimum moisture content increased before the compactive energy decreased with depth below the impact point.

Rollins and Mark (2005) described the dynamic compaction carried out in Wyoming. The deep dynamic compaction (DDC) work typically involved a 20 ton weight, 4 ft (1.22 m) in diameter, with a drop height of 100 ft (30.48 m). Generally, primary drop points were spaced at 10 and 12 ft (3.05 and 3.66 m) on centers transverse and parallel to the direction of traffic, respectively. Secondary drops were spaced at the center of four primary drop points. Five drops were typically made at each primary drop point and 2 at each secondary point. The number of drops per point was typically limited to 5 or 6 for the primary points and 2–3 for the secondary points. The applied energy per volume increased from about 60 % of the standard Proctor test energy in the 1989–1990 work to about 95 % for the second set of tests. These relatively high energy levels are common for treating collapsible soils. Inspectors monitored the number of drops and the crater depth for each drop point. The average crater depths were typically between 5 and 7 ft (1.52 and 2.13 m) deep after treatment and the crater diameter typically increased to about 10 ft (3.05 m) at the ground surface. Dynamic cone penetration tests were performed at several locations along each roadway section before and after treatment. Substantial increases in cone penetration resistance occurred when soil types consisted of silty sands (SM) or low-plasticity silts (ML or CL-ML) and the natural water content was relatively low. In these cases, the average penetration resistance increased from an average of 5–7 blows/ft to an average of 25–30 blows/ft. However, in cases where the soil profile contained layers of plastic clay (CH) with higher natural water contents, little improvement in penetration resistance was observed. Under these conditions, the clay soil appeared to absorb a significant percentage of the impact energy rather than transmitting it to the deeper layers.

Rollins and Kim (2010) presented case histories provided for 15 projects at 10 locations in the United States where collapsible soils were treated with dynamic compaction (DC). For each site, the soil properties, compaction procedures, and subsequent improvement were summarized. Although cohesionless and low-plasticity collapsible soils were successfully compacted, clay layers in the profile appeared to absorb energy and severely reduced compaction effectiveness. Correlations were presented for estimating the maximum depth of improvement, the degree of improvement versus depth, the depth of craters, and the level of vibration based on measurements made at the various sites. The compactive energy per volume was typically higher than for non-collapsible soils because collapsible soils are usually loose but relatively stiff. The maximum depth of improvement was similar to that for non-collapsible soils; however, significant scatter was observed about the best-fit line. Improvement was non-uniform with nearly 80 % of the total improvement occurring within the top 60 % of the improvement zone. The crater depth was related to a number of factors besides the drop energy including the number of drops, drop spacing, and contact pressure.

Problems with Gypseous Soils

The problems encountered in gypseous soils are summarized in the following points as reported by Saaed (1990) and Al-Abdullah (1996):

- Great losses in strength upon wetting.
- Sudden increase in compressibility upon wetting.
- Continuation of deformation and collapse upon leaching due to water movement.
- The existence of cracks due to seasonal changes.
- The existence of sink holes in the soil due to local dissolution of gypsum.

Improvement of Gypseous Soils

The improvement of gypseous soils means decreasing the effect of water on the gypseous soils to ensure the safety and stability of the engineering structures. This treatment can be achieved chemically or physically:

- The chemical treatment means that the soil properties are improved with some chemical additives, such as lime, cement, bituminous, bentonite, dehydrate calcium chloride, etc.
- The physical treatment means that the soil properties are improved using mechanical methods, such as compaction, soil reinforcement, dynamic compaction, pre-wetting, soil replacement and others.

The major purpose of this study is to make an attempt to improve gypseous soil characteristics using a dynamic compaction process. A special apparatus is manufactured to compact the soil. Different soils having different gypsum contents are used to study the effect of dynamic compaction on improving the collapsibility characteristics of these soils.

EXPERIMENTAL WORK

This section includes description of the sampling methods, the testing apparatus used in addition to the procedures of testing. Three types of natural gypsuous soils used in this study are taken from three locations of Al-Garmaa in Al-Anbar governorate western Iraq. The first and second soil samples are taken from a depth of 1.0 m and the third sample is obtained from a depth of 2.0 m below the natural ground surface. The samples are tested in the Soil Mechanics Laboratory at Al-Mustansiriya University in Baghdad. The samples were remolded to control the soil density. A detailed testing program included the main tests conducted on the samples. The testing program in this study can be summarized in the following groups:

- Classification tests are performed first including physical and chemical tests. The physical tests include specific gravity, Atterberg limits, grain size distribution and water content, while chemical tests are preformed to determine the chemical components of the samples.

- Standard Proctor and modified Proctor compaction tests are carried out to determine the moisture-density relationships.

- Compressibility test is performed using oedometer apparatus. Double oedometer tests are carried out on soaked and unsoaked samples at untreated and compacted states.

Physical Tests

The specific gravity of the soil is determined according to the British standards (BS 1377: 1990, Test No. 6 B; Head 2006), but "Kerosene" is used instead of water due to the dissolution of gypsum in water. The liquid limit test is carried out in accordance with (BS 1377: 1990, Test 2A), using cone penetrometer method. The plastic limit is determined in accordance with (BS 1377: 1990, Test No. 3). The liquid and plastic limits are carried out on soil passing sieve No. 40 and the temperature used for drying is maintained at (45–50) °C due to the presence of gypsum in the soil (ASTM 2216 1998).

The grain size distribution is determined by sieve analysis test, which is conducted in accordance with (ASTM D422 2002) with dry sieving.

The water content is performed in accordance with (BS 1377: 1990, Test (A); Head 2006). The water content is determined at drying temperature of 45 °C because the soil contains a significant amount of gypsum, thus preventing the loss of crystal water is required.

Chemical Tests

Several chemical tests are carried out on the samples. These tests included:

- Total soluble salts TSS % which are determined according to the BS 1377: (1990), Test 9.

- Total content of SO_3% which is determined according to BS 1377.

In this study, the gypsum content is determined by two methods:

Gypsum content is determined according to the British standards (BS 1377: 1990), The gypsum content is determined from the sulfate content of the soil by the following equation (BS 1377: 1990):

$$\chi\,(\%) = SO_3 \times 2.15$$

$$(1)$$

The gypsum content is also found according to the method suggested by Al-Mufty and Nashat (2000). This method consists of oven drying the soil at 45 °C until the weight of the sample becomes constant. The weight of sample at 45 °C is recorded. Then, the same sample is dried at 110 °C until the weight becomes constant and recorded.

The gypsum content is calculated according to the following equation:

$$\chi\,(\%) = \frac{W_{45°C} - W_{110°C}}{W_{45°C}} \times 4.778 \times 100$$

(2)

where χ = Gypsum content (%), $W_{45°C}$ = Weight of the sample at 45 °C, $W_{110\,°C}$ = Weight of the sample at 110 °C.

Mechanical Tests

Compaction Tests

Standard and modified compaction tests are carried out for the untreated soils to determine the moisture-unit weight relationship.

- Standard Compaction Test: This test is performed according to (ASTM D698, Method A2003).
- Modified Compaction Test: This test is carried out according to (ASTM D1557, Method A2003).

Compressibility Tests

A series of oedometer tests is performed in accordance with (ASTM 2435 2002). These tests are carried out on untreated and compacted samples to determine the compressibility characteristics. To conduct the tests, the fixed type consolidometer cells and loading frame with specimens of 75 mm diameter and 20 mm height are used.

Double Oedometer Test

This test is conducted according to the procedure suggested by Jennings and Knight (1957). In this test, two identical samples are tested independently. The first one is loaded without the addition of any water (unsoaked). Another sample is soaked and then loaded progressively as in the standard consolidation test. Both samples are stressed beginning from (25) kPa. Then, the test is continued following the standard procedure of doubling the applied loads until a stress of 800 kPa is reached. The sample is then unloaded by stress decrements through two stages 400 and 200 kPa. The difference between the two compression curves quantifies the amount of deformation that would occur at any stress level if the soil to be saturated during its loading history.

Description of Dynamic Compaction Apparatus

The full details of the soil dynamic compaction apparatus can be shown in Figure. 1. All components of apparatus are made of rigid steel.

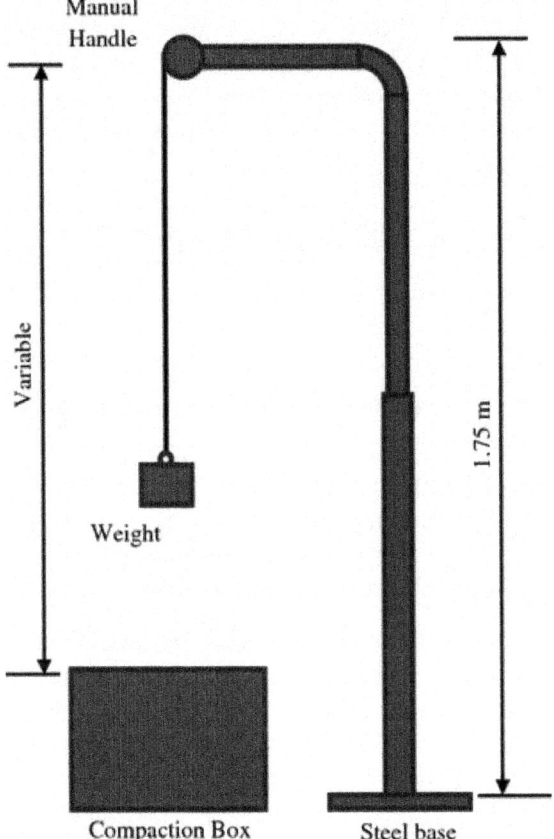

Figure.1: Details of soil dynamic compaction apparatus parts.

Generally, the apparatus consists of:

The Compaction Box

It is a steel box of $(50 \times 50 \times 35)$ cm dimensions which has a gate in order to make the operation of extrusion of the compacted samples easy. The box can be moved by a group of rollers.

Compaction Arm

It is a steel pipe having the form of letter (L) connected from the bottom with a steel base of dimensions (36 × 36) cm. The side part of compaction arm is a steel pipe of length L = 40 cm and diameter D = 9 cm, it ends with a toothed crow bar rotated manually by hand for controlling the height of falling of weights and to ensure free fall of the weights. This is done through controlling length of the metal wire, with which it is connected during the movement of the toothed crow bar of other side. The middle part of the compaction arm is interfered pipe of steel. The outer pipe which is connected to the base of the arm has dimensions of inner diameter ID = 10 cm, outer diameter OD = 11 cm and height H = 100 cm. The interior pipe which is connected to the side part has dimensions of D = 9 cm and H = 75 cm.

PREPARATION OF THE SOIL SAMPLE

In this study, compacted soil samples are prepared at moisture content equal to the optimum moisture content according to ASTM D698 (2003). The preparation of the soil sample is summarized in the following:

- The natural water content is determined according to (BS 1377: 1990, test A; Head 2006).
- Dry soil is mixed carefully with the required amount of water till it reaches the optimum moisture content.

Soil Compaction by Dynamic Compaction Apparatus

The compaction box is filled with soil. The soil layer thickness in the box is limited to 20 cm due to sample preparation requirements. It is known from previous studies on dynamic compaction, such as Rollins and Mark (2005), that for the used loads, 2.5–4 cm is the effective depth of influence of dynamic compaction, from which it is sufficient to obtain samples (after dynamic compaction) for oedometer test. The soil is compacted through the falling weights at different heights and the required number of blows. In this study, three weights of 2, 3 and 5 kg are used with three heights of falling 35, 50 and 65 cm and three different numbers of blows; 20, 30 and 40 are tried.

The required weight is fastened by the metal wire and raised to the specific height by moving the hold by hand then it is locked with the toothed crow bar.

The lock of the crow bar is opened, and then the weight falls freely on the soil. This operation is repeated to reach the required number of blows. The loads are controlled to be distributed at enough area for extraction of at least two samples of the compacted soil which are necessary for preparing samples for oedometer test. The box is to be rolled by hand to ensure distribution of the blows over the required area.

Two rings of the oedometer test apparatus are pushed in the compacted space of the soil to get the required samples for oedometer test.

The gate of the compaction box is opened to extract the samples.

RESULTS AND DISCUSSION

Grain Size Distribution

The results of particle size distribution tests conducted using dry sieve analysis are shown in Figure. 2. From these results, the soil specimens S_1, S_2 and S_3 can be classified according to the Unified Soil Classification System (USCS) as poorly graded—silty sand (SP − SM). It should be noticed that the classification of gypseous soil depends on the state of testing or method of calculation. Some researchers pretreated the gypseous soil with a solution such as EDTA (Ethylenediaminetetraacetic acid) (Seleam 1988) or large amount of distilled water to remove the gypsum prior to carrying out the classification tests, while others used kerosene or white spirit to prevent any more dissolution of gypsum from the soil sample.

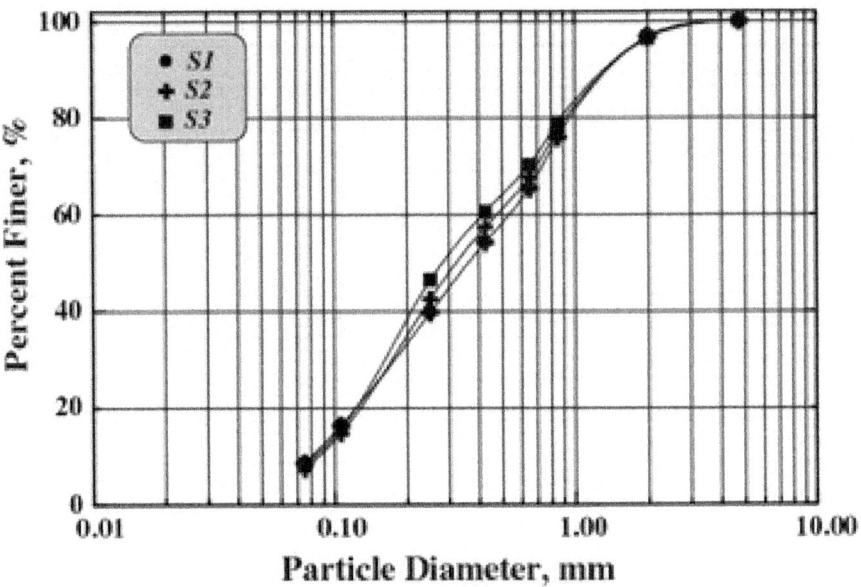

Figure.2: Grain size distribution curves for the three soils.

Specific Gravity

It can be noticed that the specific gravity decreases for the soil having high gypsum content. The low specific gravity of the three soils, which ranges between 2.28 and 2.41 is attributed to the low specific gravity of gypsum, which is equal to 2.32, compared to other soil constituents.

Atterberg Limits

Atterberg limits tend to increase with the increase of gypsum content (Al-Gabri 2003). This behaviour may be related to the small particles of gypsum, which cause an increase in the surface area of the soil, the requirements of water are increased until these limits. Similar results are found in this study as can be noticed in Table 1 in which the liquid limit values are high despite that the soil is granular. This may be attributed to the fact that the soil has apparent cohesion caused by the presence of gypsum which gives the soil the consistency of fine soil. The results of Atterberg limits are summarized in Table 1.

Table 1: Summary of physical properties of the three soils

Soil property	Type of soil		
	S_1	S_2	S_3
Specific gravity, Gs	2.28	2.34	2.41
Initial void ratio, e_o	0.915	0.85	0.927
Initial water content (%)	0.91	0.82	0.73
Liquid limit, L.L (%)	55	43	41
Plastic limit, P.L (%)	39	31	29
Plasticity index, PI (%)	16	12	12
Coefficient of curvature, C_c	0.842	0.73	1.01
Uniformity coefficient, C_u	6.42	5.17	4.17
Percent of fines (%)	9	7	8
Sand percent (%)	91	93	92
Soil classification according to (USCS)	SP – SM	SP – SM	SP – SM

Chemical Tests

Table 2 shows the results of chemical tests. Since the gypsum content of the three soils is more than (25 %); the soils are classified according to Barzanji (1973) as "highly gypsiferous".

Table 2: Results of chemical tests

Type of soil	Total soluble salts TSS (%)	BS 1377		Al-Mufty and Nashat (2000)	
		SO_3(%)	Gypsum content, χ (%)	SO_3 (%)	χ (%)
S_1	67.83	28.84	60.5	27.3	58.7
S_2	50.63	20.87	41.1	18.84	40.5
S_3	30.51	13.32	27	12	25.8

Compaction Testes

The results of compaction tests are tabulated in Table 3. The relationships between dry unit weight and water content for the tested soils are shown in Figure. 3 for compactive efforts associated with the modified and standard Proctor test.

Table 3: Results of compaction tests

Type of test	Soil property	Type of soil		
		S_1	S_2	S_3
Standard compaction	Maximum dry unit weight, kN/m³	14.45	15.35	15.2
	Optimum water content (%)	14.5	15.0	15.0
Modified compaction	Maximum dry unit weight, kN/m³	19.67	18.5	16.5
	Optimum moisture content (%)	10.0	12.7	13.8

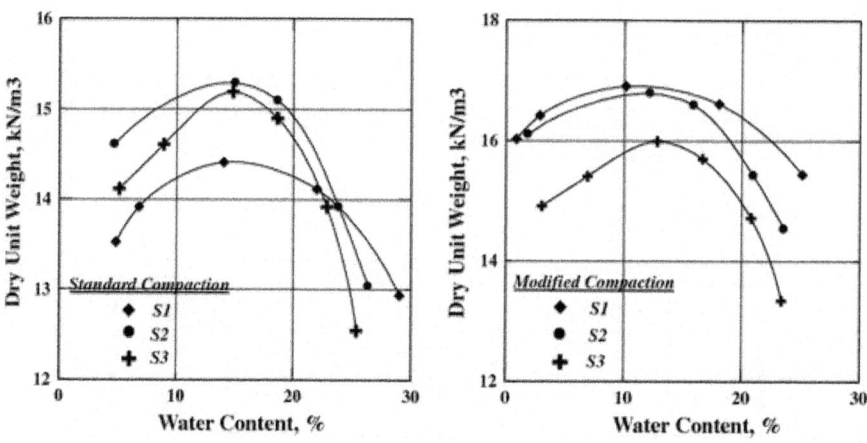

Figure.3: Compaction curves for soils S_1, S_2 and S_3.

It is noticed that the standard maximum dry unit weight of soil S_2 is somewhat higher than the standard maximum dry unit weight of soil S_1, while the opposite is true for the modified compaction test. This is due to the role of gypsum in the compaction (Al-Mufty 1997): First, the gypsum particles act as pore filling fines especially if the gypsum particles are of small size compared to the soil grains, thus, increasing the dry unit weight. Second, gypsum causes cementation to soil particles that helps resistance to compaction effort and increases the required water content to reach the maximum unit weight. In other words, the decrease in the maximum dry unit weight may be attributed to the loss of some compactive energy in breaking the cementation bonds, which may form between clay and gypsum particles. Third, increasing gypsum at the expense of soil particles causes a lower specific gravity for the soil as a whole, and leads to decreasing the dry unit weight.

It can be concluded from the water content—unit weight relationships that the test results depend on the soluble salt content (gypsum content), water content, the solubility degree of gypsum in water and compactive effort. This agrees with the findings of Farwana and Majid (1988).

Compression Test

This test is carried out on two samples for each soil. The first sample is tested at the optimum water content, while the other is directly tested after soaking in water. The results are presented as void ratio versus logarithm of effective stress as shown in Figures. 4, 5, 6, 7, 8, 9, 10, 11, 12, 13, 14, 15, 16 for all tested specimens. Figureure 4 presents the relationship for the untreated samples, while Figures. 5, 6, 7, 8, 9, 10, 11, 12,13, 14, 15, 16 show the relationship for samples tested after treatment by dynamic compaction under different numbers of blows, falling weights and height of falling of the weights.

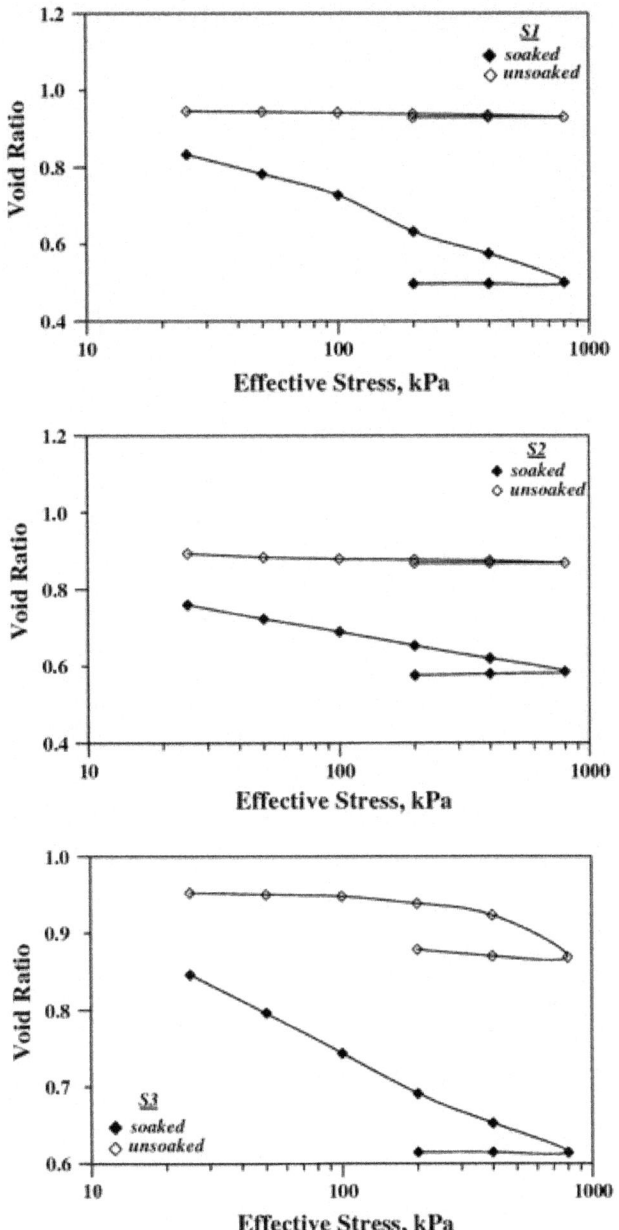

Figure.4: Results of compression test on untreated soils.

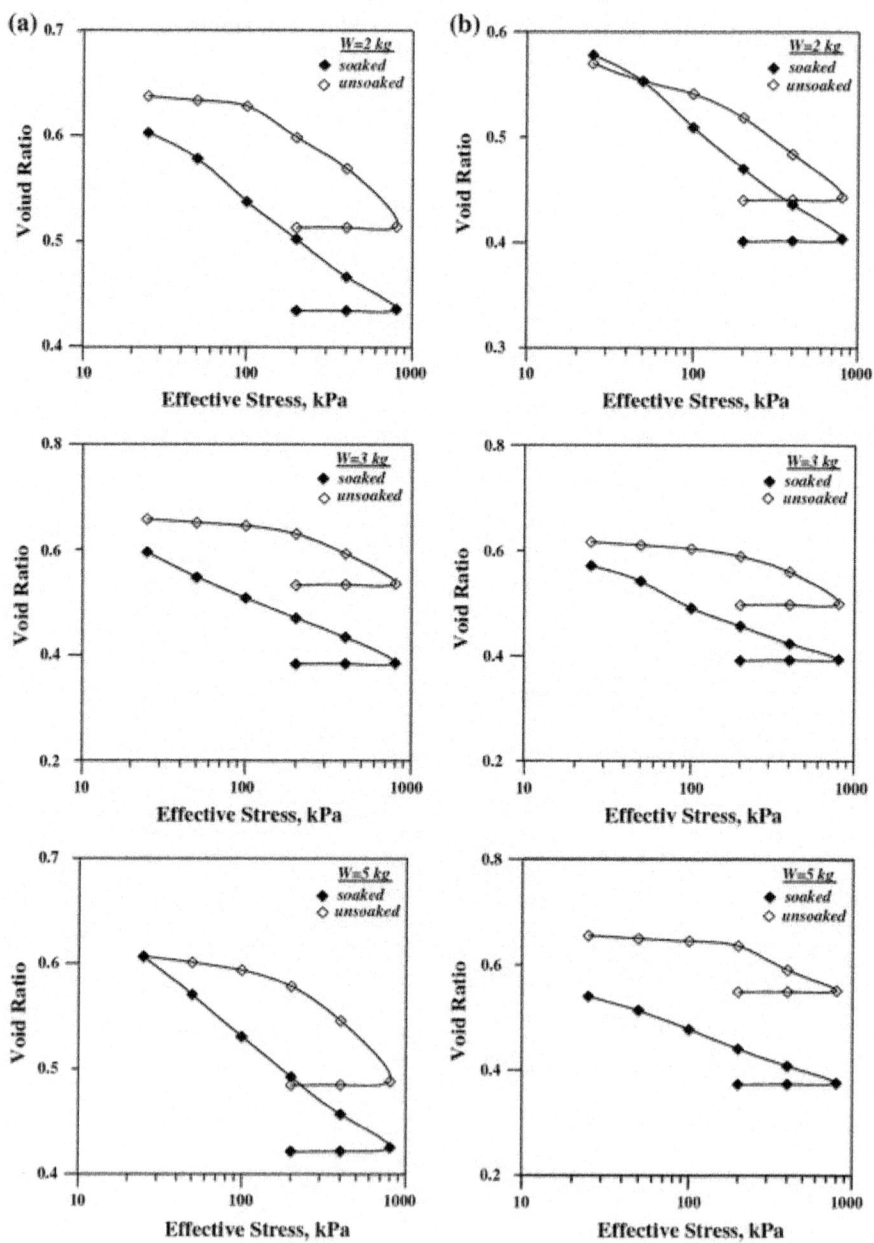

Figure.5: Results of compression test on soil S_1 compacted at no. of blows 20 and height of drop: **a** H = 35 cm, **b**H = 50 cm.

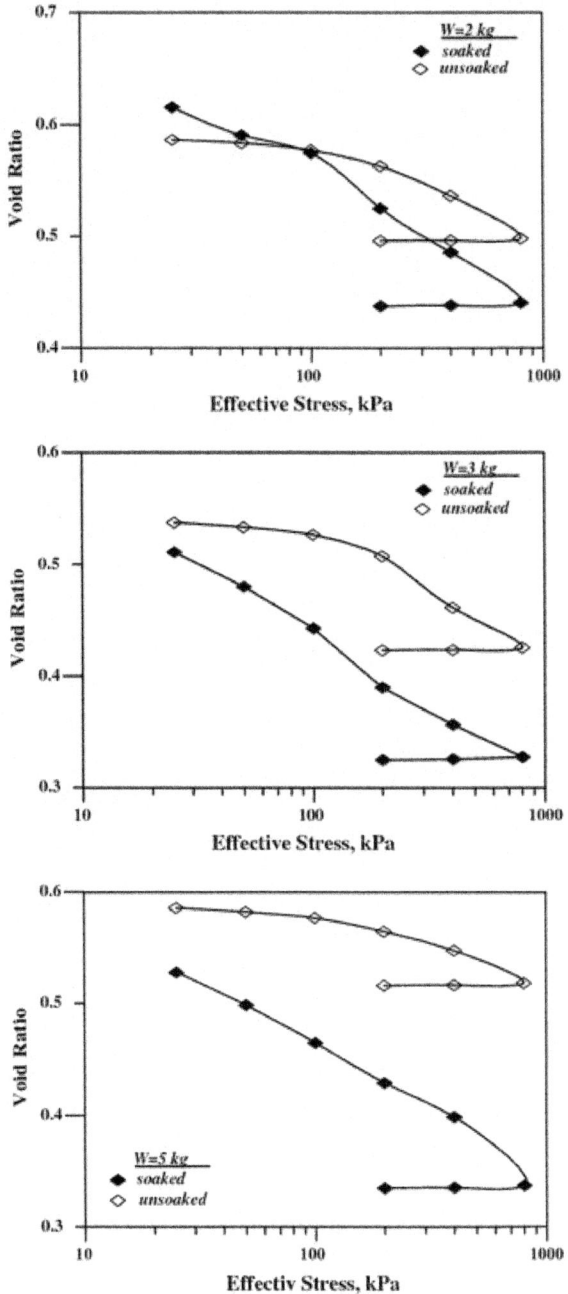

Figure.6: Results of compression test on soil S_1 compacted at no. of blows 20 and height of drop (H = 65) cm.

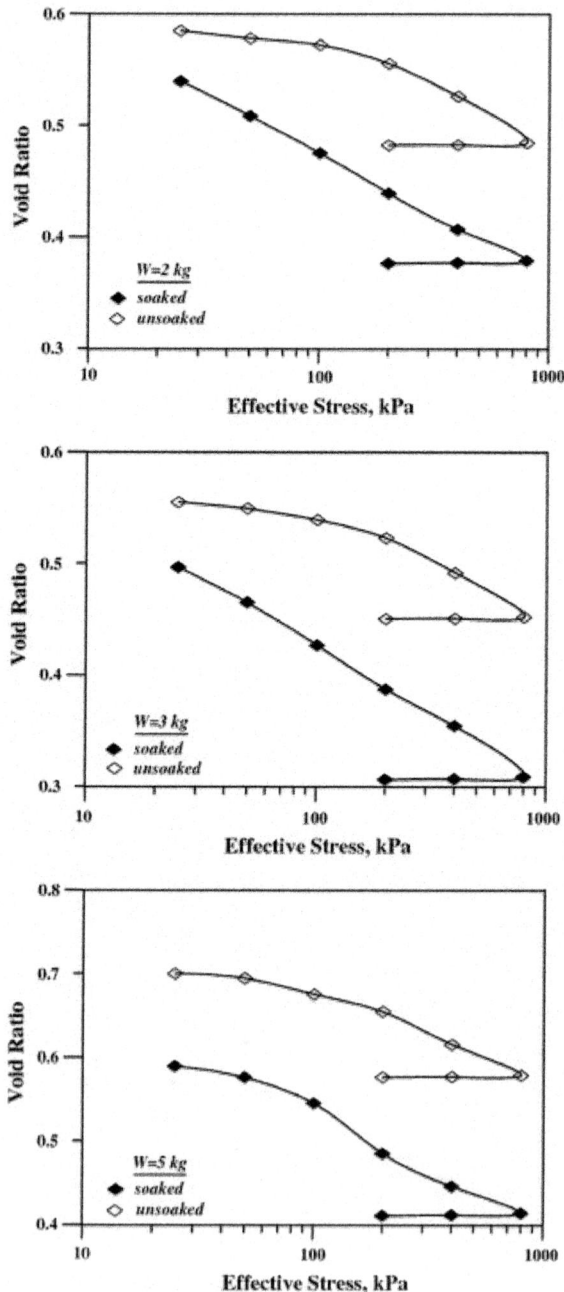

Figure.7: Results of compression test on soil S$_1$ compacted at no. of blows 30 and height of drop (H = 35) cm.

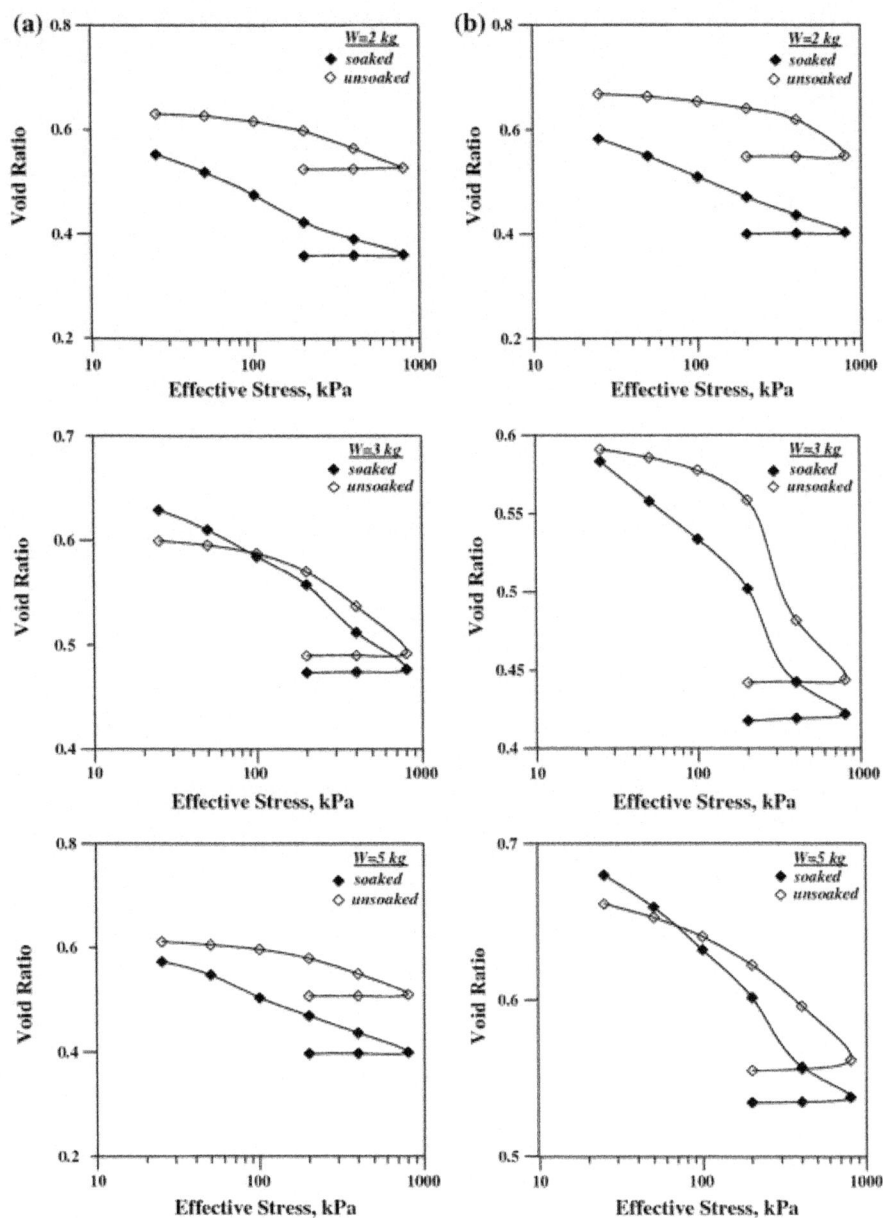

Figure.8: Results of compression test on soil S_1 compacted at no. of blows 30 and height of drop: **a** H = 50 cm, **b**H = 65 cm

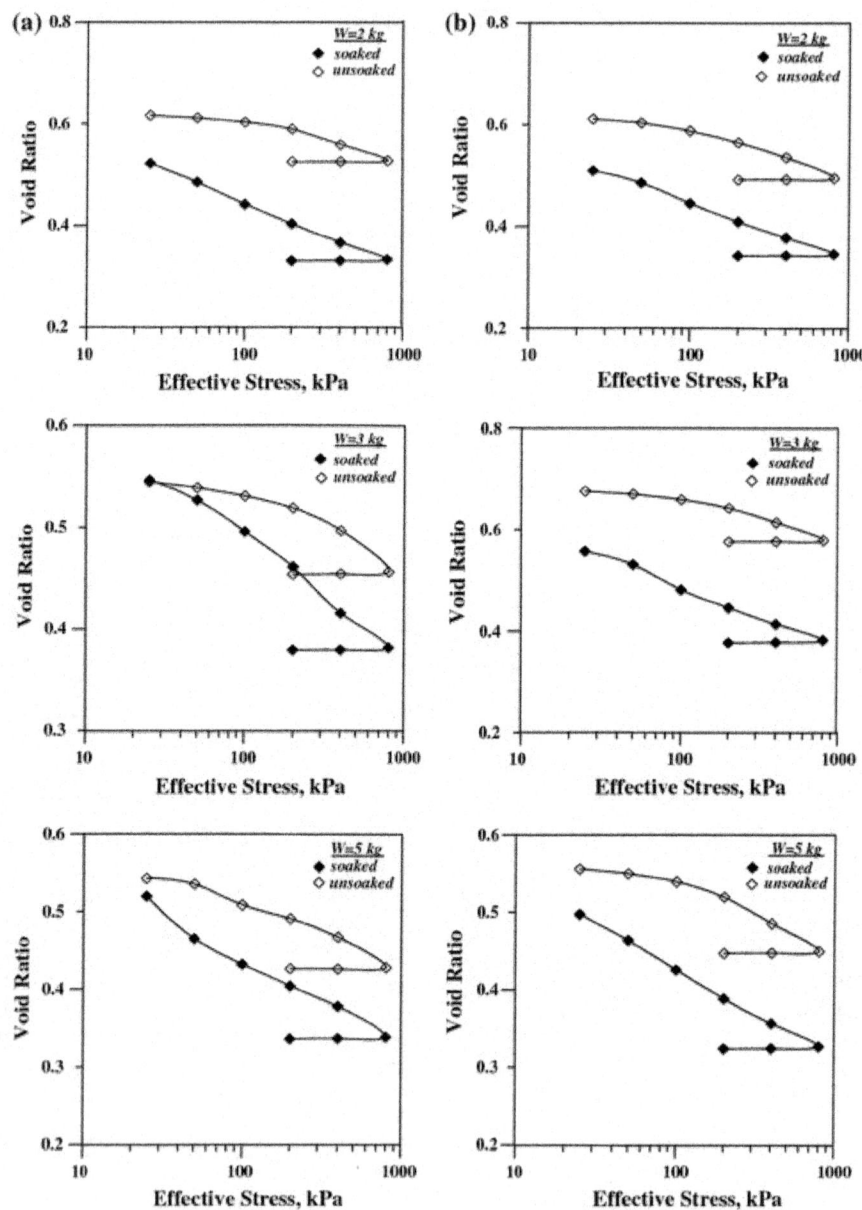

Figure.9: Results of compression test on soil S_1 compacted at no. of blows 40 and height of drop: **a** H = 35 cm, **b** H = 50 cm.

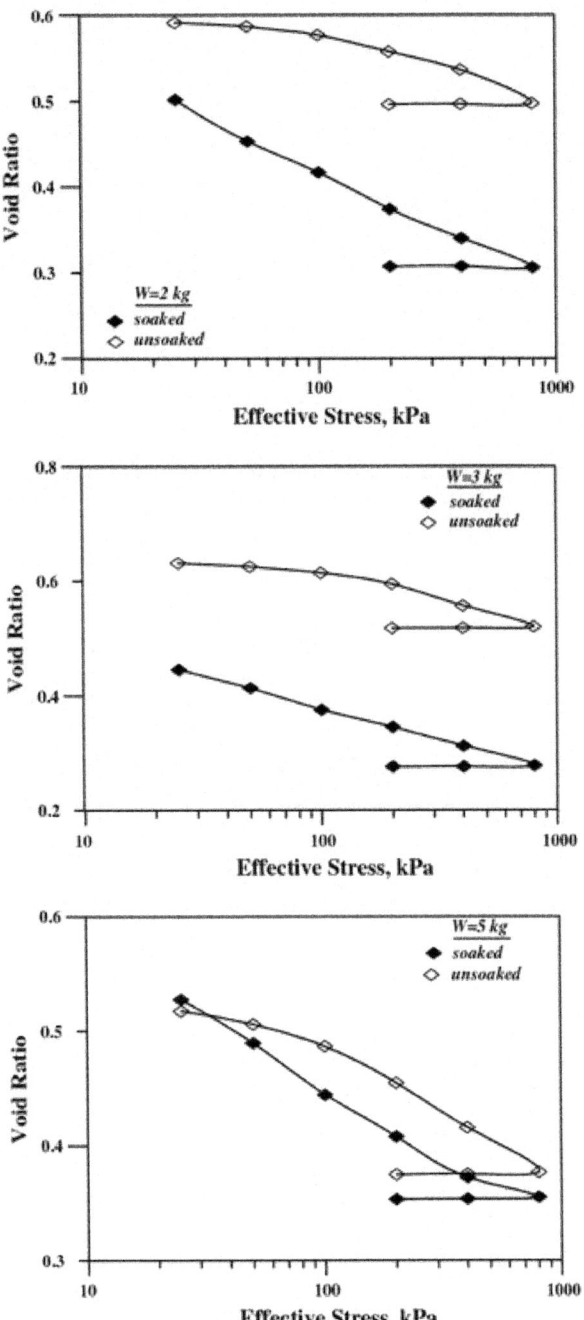

Figure.10: Results of compression test on soil S_1 compacted at no. of blows 40 and height of drop (H = 65) cm.

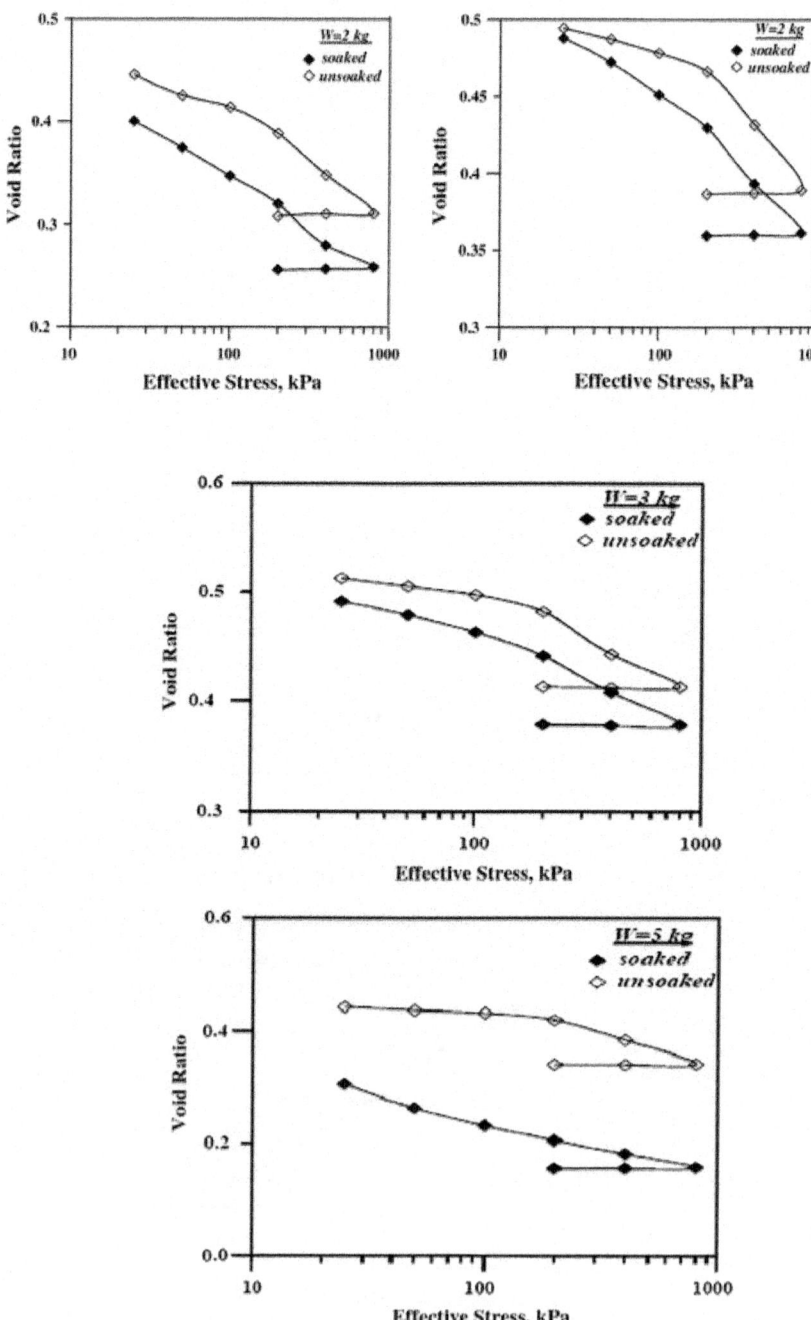

Figure.11: Results of compression test on soil S$_2$ compacted at no. of blows 20 and height of drop (H = 35) cm.

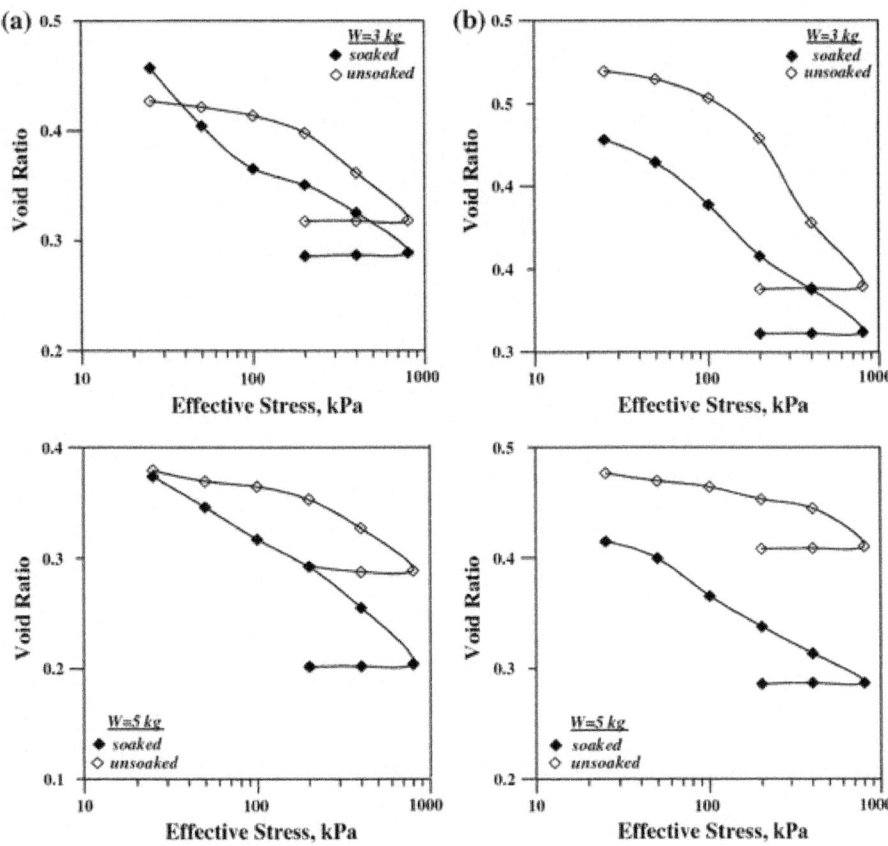

Figure.12: Results of compression test on soil S_2 compacted at no. of blows 20 and height of drop: **a** H = 50 cm, **b**H = 65 cm.

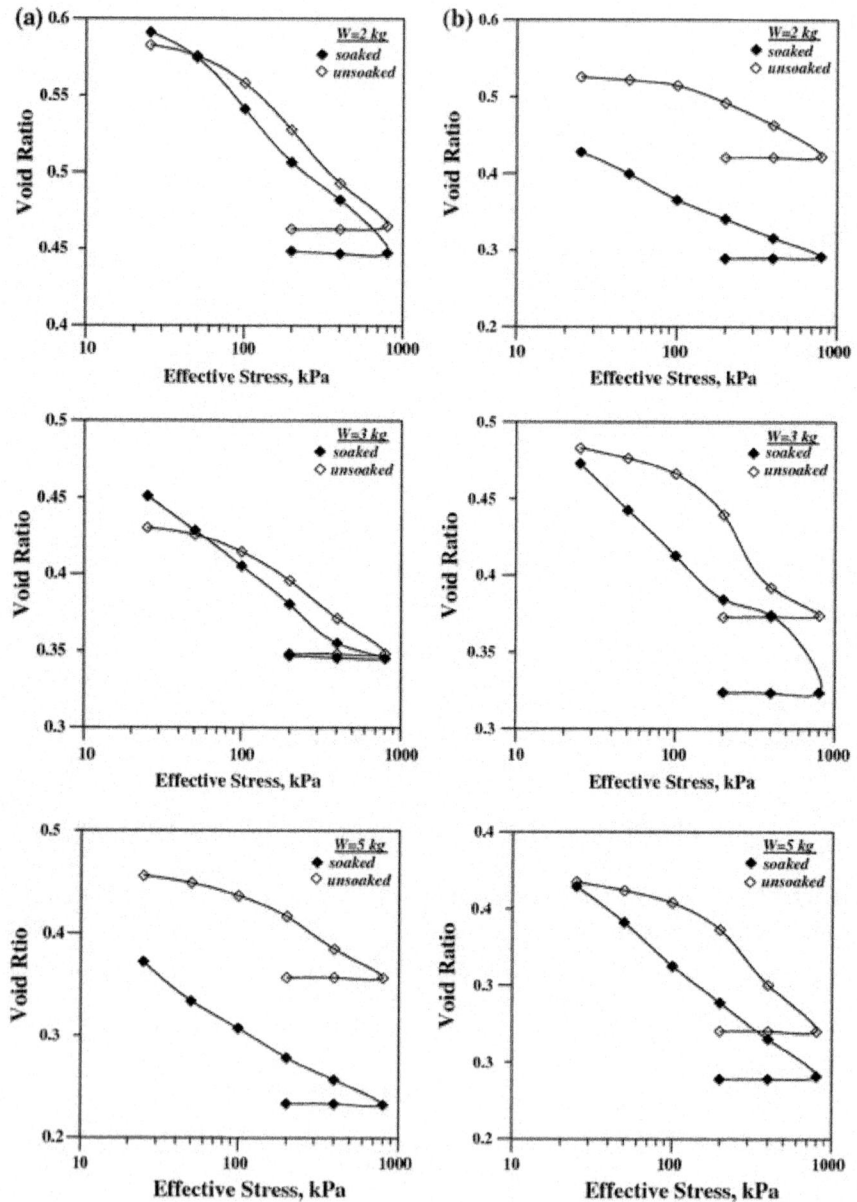

Figure.13: Results of compression test on soil S_2 compacted at no. of blows 30 and height of drop: **a** H = 35 cm, **b**H = 50 cm

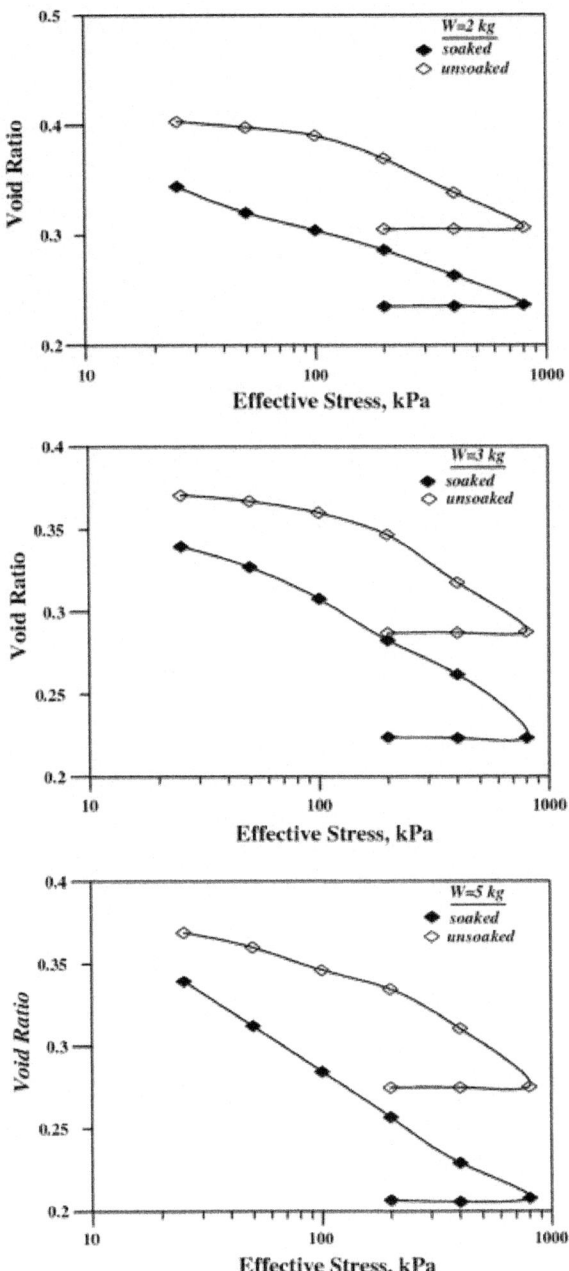

Figure.14: Results of compression test on soil S$_2$ compacted at no. of blows 30 and height of drop (H = 65) cm

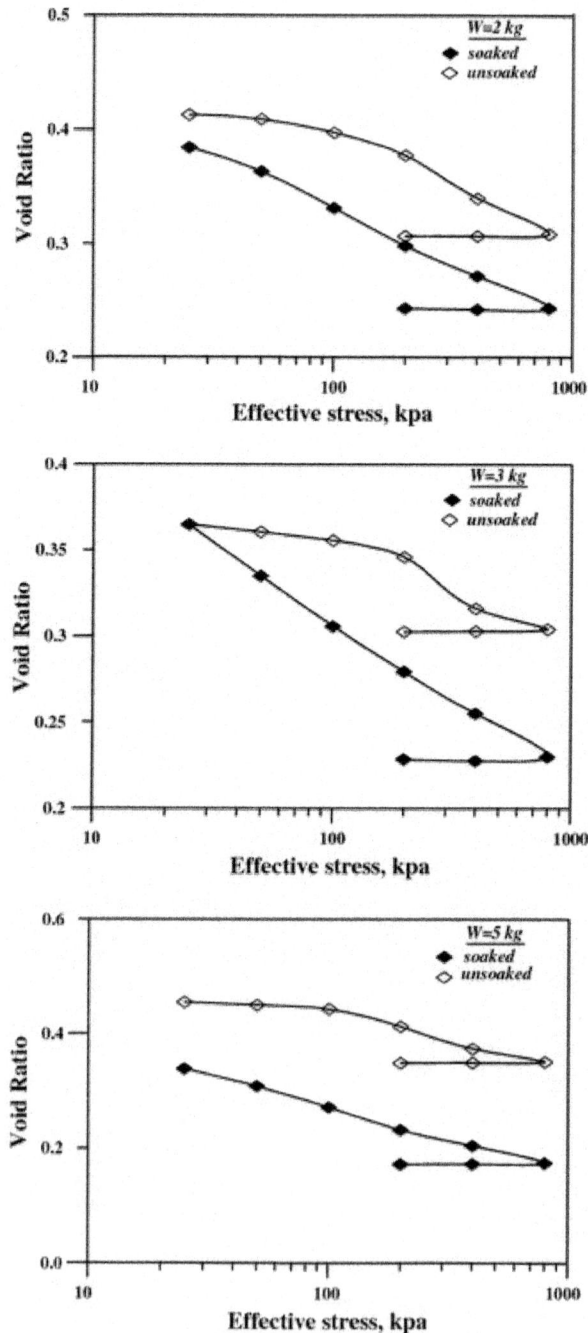

Figure.15: Results of compression test on soil S_2 compacted at no. of blows 40 and height of drop (H = 35) cm.

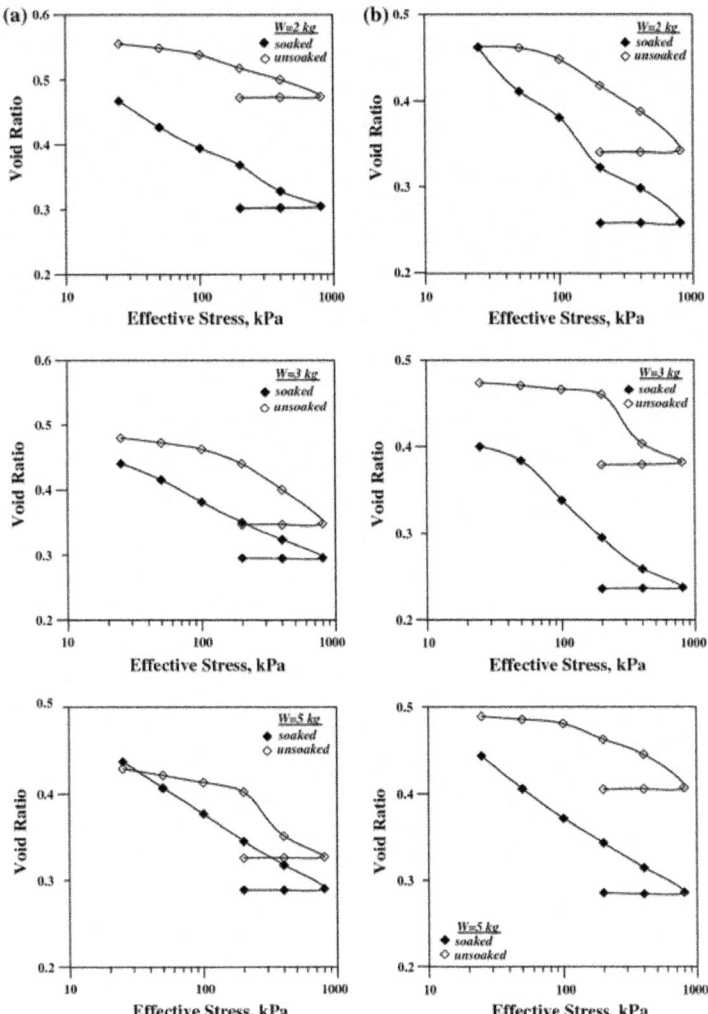

Figure.16: Results of compression test on soil S_2 compacted at no. of blows 40 and height of drop: **a** H = 50 cm, **b**H = 65 cm.

From these results, it can be seen that:

- The initial void ratio e_o for the compacted sample of soil S_1 is greater than the initial void ratio of soils S_2 and S_3. This is due to the effect of gypsum content. The presence of gypsum enlarges the voids within the soil structure.

- The void ratio at stress level 800 kPa increases slightly with the increase in the height of drop H from 35 cm to 50 cm, but decreases under H = 65 cm.

- The void ratio at stress level 800 kPa for soaked compacted samples decreases with the increase in the falling weight.

- The void ratio at stress level 800 kPa for soaked compacted samples decreases with the increase in the number of blows.

- In samples subjected to dynamic compaction, the change in void ratio Δe upon soaking becomes smaller than that of untreated samples which means that the collapse potential decreases.

In Figures. 17, 18, 19, the compression index C_c is plotted versus the number of blows. It becomes clear that the compression index for soil S_1 increases from 0.011 to 0.221 upon soaking in water. The same behaviour is observed for the other two soils, as shown in Table 4. This behaviour is attributed to the fact that the gypsum acts as cementing material by adding resistance to deformation, but upon wetting, softening occurs due to cement loss.

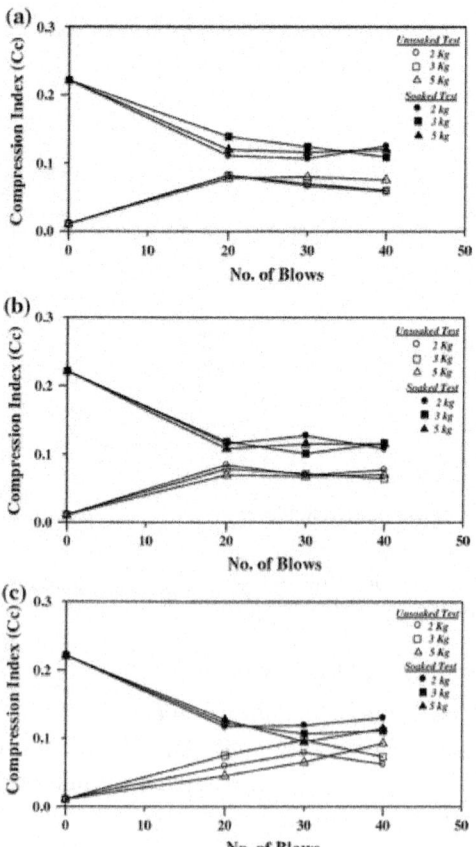

Figure.17: Number of blows versus compression index for compacted soil S_1 at: **a** H = 35 cm, **b** H = 50 cm and **c** H = 65 cm.

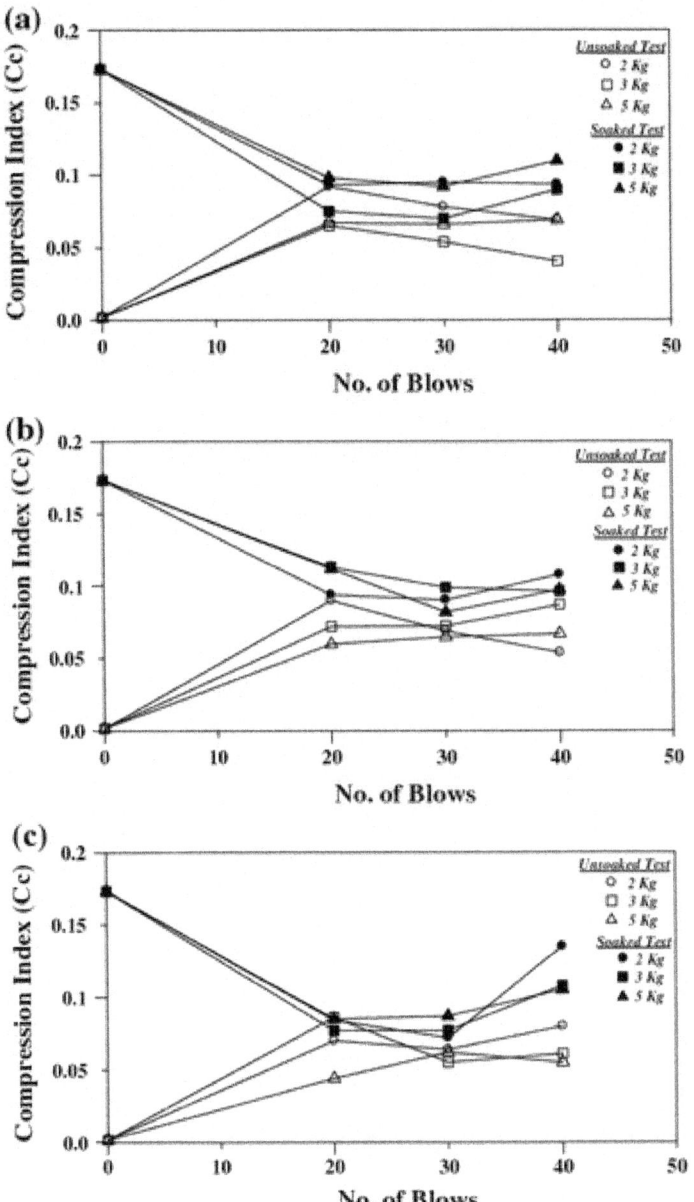

Figure.18: Number of blows versus compression index for compacted soil S_2 at: **a** H = 35 cm, **b** H = 50 cm and **c** H = 65 cm.

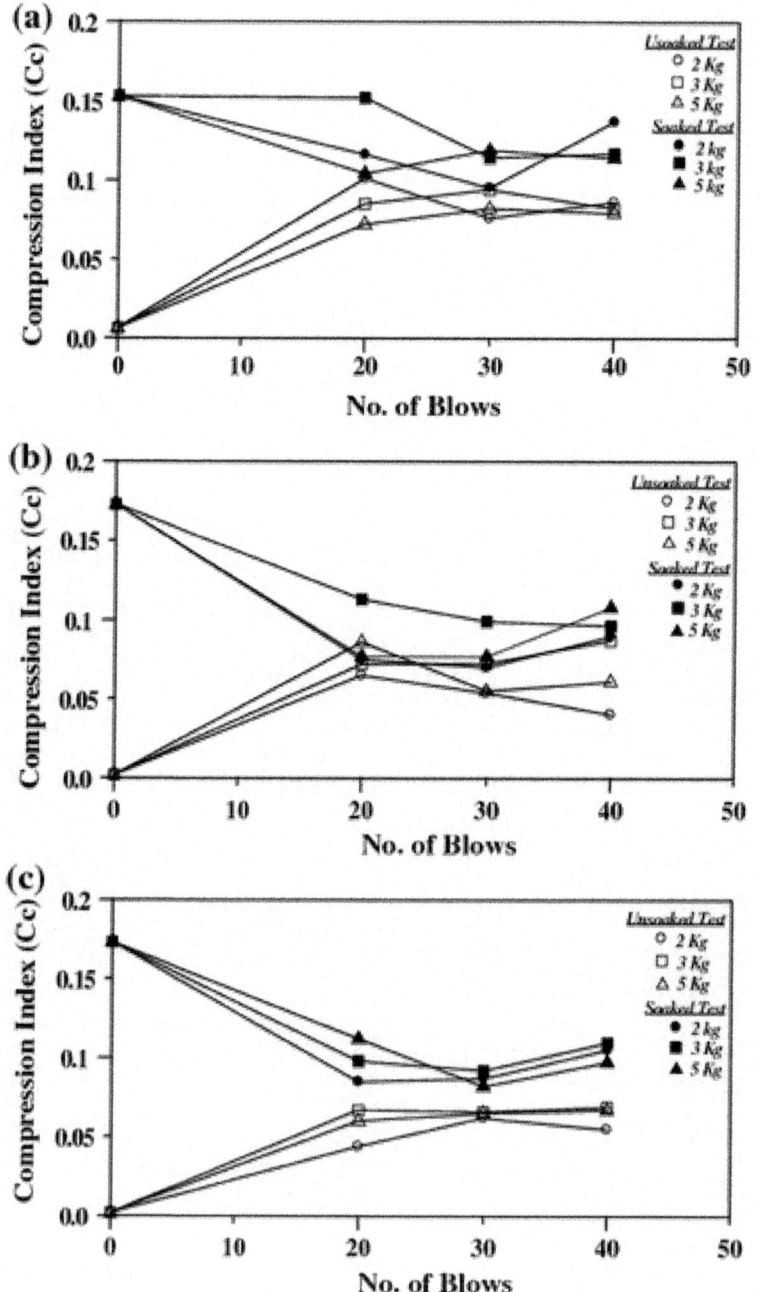

Figure.19: Number of blows versus compression index for compacted soil S_3 at: **a** H = 35 cm, **b** H = 50 cm and **c** H = 65 cm.

Table 4: Summary of compression test results of untreated soils

Type of test	Type of soil	C_c	C_r
Soaked	S_1	0.221	0.0054
	S_2	0.173	0.00163
	S_3	0.153	0.0016
Unsoaked	S_1	0.011	0.00016
	S_2	0.00163	0.00017
	S_3	0.00152	0.00083

It can be also noticed that, for saturated soil samples, the compression index decreases as the sample is compacted under a number of blows among 20-40. The best improvement in compressibility is achieved when the sample is subjected to 20 drops. Above this number, a negligible decrease in C_c is obtained. For soil S_1, as the number of blows increases, the C_c increases due to the increase in the height of drop, while for soils S_2 and S_3, the large number of blows greater than 30 may lead to increase C_c. This may be attributed to breakage of soil grains and a decrease in particle interlocking and hence increase in void ratio. The decrease in compression index ranges between 0.152 and 0.044 %, while soil S_1 shows inverse behaviour. This behaviour is attributed to the effect of gypsum content.

The dry samples show inverse behaviour; the compression indices increase when samples are compacted under 20 blows, then this increase becomes smaller when the number of blows increases further. The reason for this can also be attributed to breaking of particles which in turn causes an increase in void ratio.

The height of drop has a noticeable effect on C_c values. As the height of drop increases from 35 cm to 65 cm, the compression index decreases. This effect is greater in samples S_1 and S_2 than sample S_3. As the gypsum content increases, the dynamic compaction has greater effect on the improvement of compressibility of the soil. It can be seen that the best improvement in C_c takes place for soil S_2 where the gypsum content, $\chi = 41.1$ %, after which, this gain decreases.

The same behaviour can be seen in Figures. 20, 21, 22, which show the relationship between the rebound index C_r and the number of blows for samples compacted at different weights and heights of drop. Generally, from these Figureures, it can be noticed that for soaked soil sample S_1, C_r increases with increasing the number of blows, while for the unsoaked sample, C_r decreases at a number of blows equals to 30 then it increased at a number of blows equals to 40.

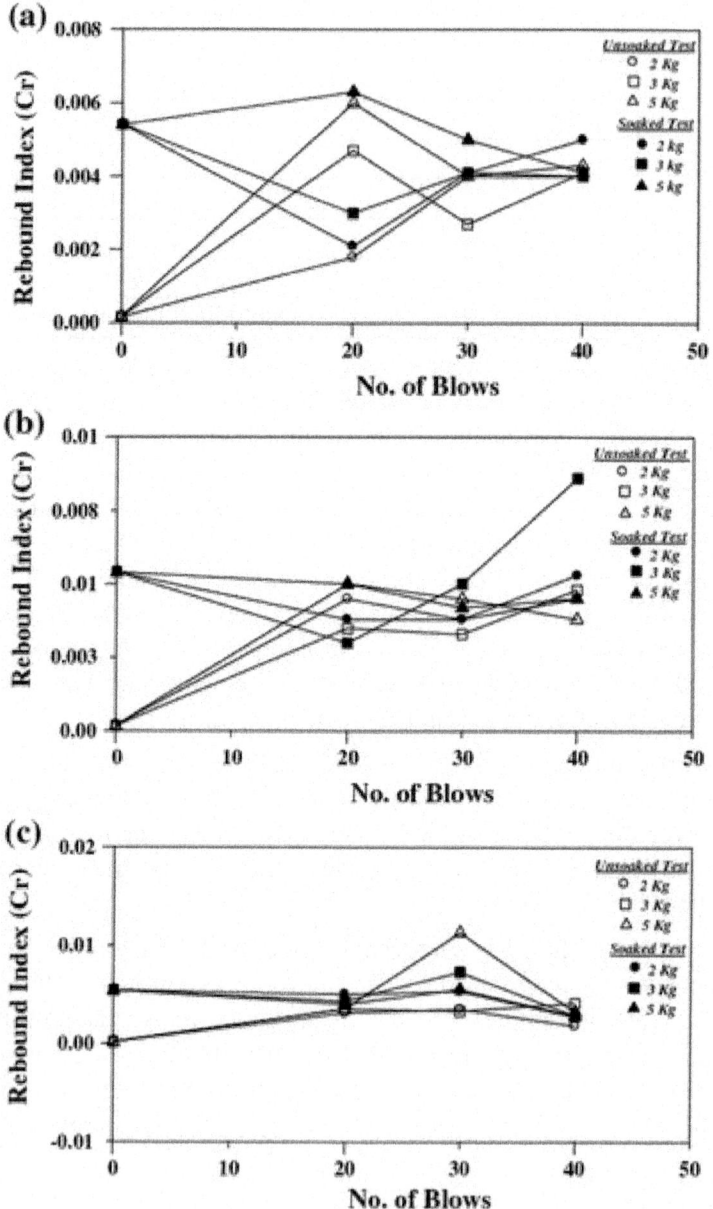

Figure.20: Number of blows versus rebound index for compacted soil S_1 at: **a** H = 35 cm, **b** H = 50 cm and **c** H = 65 cm.

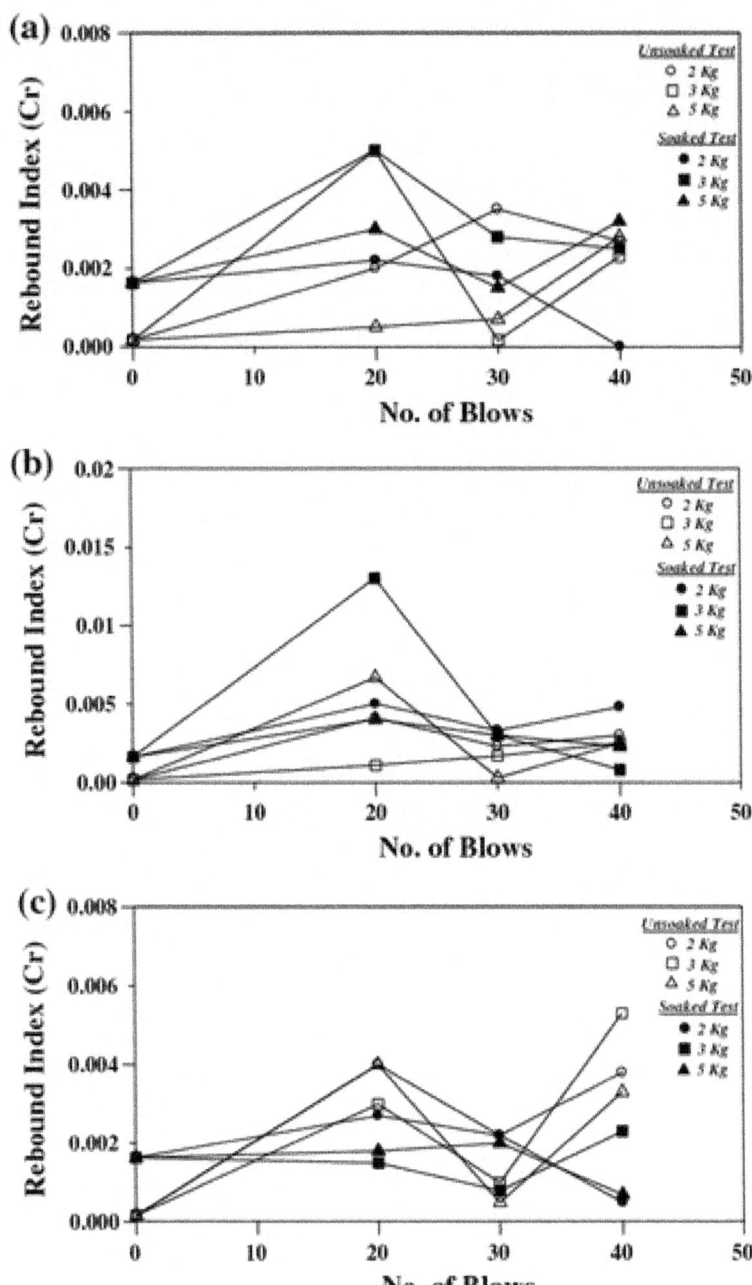

Figure. 21: Number of blows versus rebound index for compacted soil S_2 at: **a** H = 35 cm, **b** H = 50 cm and **c** H = 65 cm.

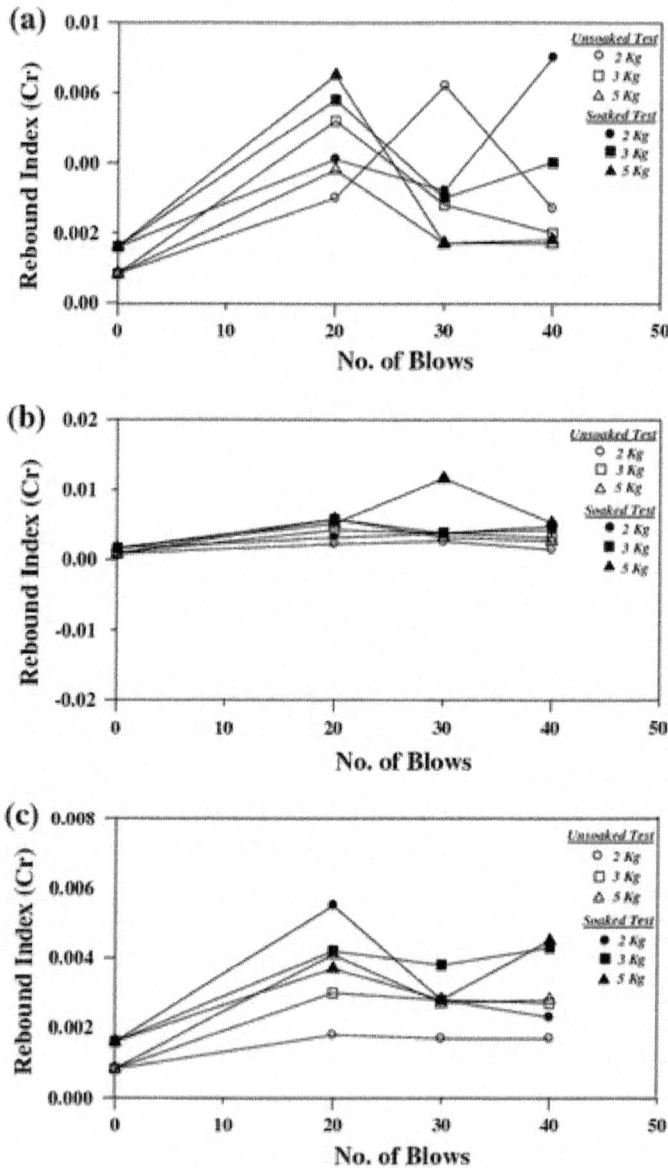

Figure.22: Number of blows versus rebound index for compacted soil S₃ at: **a** H = 35 cm, **b** H = 50 cm and **c**H = 65 cm.

For soil S_2, it can be observed that for soaked samples, C_r decreases with increasing the number of blows, while for the unsoaked samples, C_r decreases at a number of blows equals to 30, then it increased at a number of blows equals to 40. The opposite is true for soil S_3.

The improvement in C_r decreases with the increase in gypsum content. The improvement is greater for sample S_3 where $\chi = 27$ %. This may be attributed to the presence of gypsum fragments within these soils, whose crystallization structures have extremely strong structural bonds, but are very brittle and are irreversibly destroyed by mechanical action during the loading stage (Petrukhin and Arakelyan 1985; Al-Qaissy 1989).

CONCLUSIONS

- In this work, three samples with different gypsum contents are treated by both standard compaction tests and dynamic compaction apparatus. From the results and analysis of the tests on soil samples with gypsum content 60.5, 41.1 and 27 %, the following conclusions can be made:

- The best improvement in compressibility is achieved when the samples are subjected to 20 drops, this conclusion is based on the improvement of compression index of soaked samples obtained after treatment. Above this number, a negligible decrease in the compression index C_C is obtained.

- As the height of drop increases from 35 to 65 cm, the compression index C_C decreases. This effect increases with the increase in the gypsum content. The same behaviour of the compression index C_C can be noticed on the rebound index C_r.

- As the gypsum content increases, the dynamic compaction has greater effect on improvement of compressibility of the soil. In samples subjected to dynamic compaction, the change in void ratio Δe upon soaking becomes smaller than that of untreated samples which means that the collapse potential decreases.

REFERENCES

1. Al-Abdullah SFI (1996) The upper limit of gypseous salts in the clay core of Al-Adhaim Dam. Ph.D. Thesis, Department of Civil Engineering, University of Baghdad, Iraq

2. Al-Gabri MKA (2003) Collapsibility of gypseous soils using three different methods. M.Sc. Thesis, Building and Construction Engineering Department, University of Technology, Baghdad, Iraq

3. Al-Mufty AA (1997) Effect of gypsum dissolution on the mechanical behavior of gypseous soils. Ph.D. Thesis, Civil Engineering Department, University of Baghdad, Iraq

4. Al-Mufty AA, Nashat EH (2000) Gypsum content determination in gypseous soils and rocks. In: Proceedings of the 3th Jordanian

international mining conference, Amman, vol 2, pp 485–492

5. Al-Qaissy FF (1989) Effect of gypsum content and its migration on compressibility and shear strength of the soil. M.Sc. Thesis, Building and Construction Engineering Department, University of Technology, Baghdad, Iraq

6. ASTM D2216 (1998) Standard test method for laboratory determination of water (moisture) content of soil and rock by mass. Annual book of ASTM standards, vol 04.08, ASTM international, West Conshohocken, PA, pp 1–5

7. ASTM D422 (2002) Standard test method for particle size analysis of soils. Annual book of ASTM standards, vol 04.08, ASTM International, West Conshohocken, PA, pp 1–8

8. ASTM D2435 (2002) Standard test method for one-dimensional consolidation properties of soils. Annual book of ASTM standards, vol 04.08, ASTM international, West Conshohocken, PA, pp 1–10

9. ASTM D698 (2003) Standard test method for laboratory compaction characteristics of soil using standard effort. Annual book of ASTM standards, vol 04.08, ASTM international, West Conshohocken, PA, pp 1–8

10. ASTM D1557 (2003) Test methods for laboratory compaction characteristics of soil using modified efforts. Annual book of ASTM standards, vol 04.08, ASTM international, West Conshohocken, PA, pp 1–8

11. Barzanji AF (1973) Gypesous soil of Iraq. Ph.D. Thesis, University of Ghent, Belgium

12. BS 1377 - 2 (British Standard Institution) (1990) Method of test for soils for civil engineering purposes. British Standard Institution, London, UK

13. FAO (1990) Management of gypsiferous soils. FAO Soils Bull 62

14. Farwana TA, Majid ZK (1988) An investigation into the use of emulsified asphalt in the stabilization of sandy sabkha. In: Proceedings of the 3rd IRF Middle East, Regional Meeting, Kingdom of Saudi Arabia

15. Head KH (2006) Manual of soil laboratory testing, Vol. 1, Soil classification and compaction tests, 3rd edn. Whittles publishing, UK

16. Jennings JK, Knight K (1957) The additional settlement of foundation due to a collapse of structure of sandy subsoil on wetting. In: Proceedings of 4th international conference on soil mechanics and foundation engineering, vol 2, pp 316–319

17. Nashat IH (1994) Exploit of gypseous soil. Report, Ministry of Housing and Construction—National Center for Construction Laboratories NCCL, Baghdad, Iraq, pp 54–63

18. Nettleton WD, Nelson RE, Brasher BR, Derr PS (1982) Gypsiferous soils in the Wester United States. In: Soil Science Society of America Special Publication, No. 10, pp 147–168

19. Petrukhin VP, Arakelyan EA (1985) Strength of gypsum-day soils and its variation during the leaching of salts. J Soil Mech Found Eng 21(6):23–25

20. Rollins KM, Kim J (2010) Dynamic compaction of collapsible soils based on U.S. case histories. J Geotech Geoenviron Eng ASCE 136(9):1178–1186

21. Rollins KM, Mark F (2005) Dynamic deep compaction treatment of collapsible soils in Wyoming. Salt Lake City annual meeting, October 16–19, 2005, Geological Society of America Abstracts with Programs, vol 37, No. 7, p 328

22. Rollins KM, Jorgensen ST, Ross TE (1998) Optimum moisture content for dynamic compaction of collapsible soils. J Geotech Geoenviron Eng ASCE 124(8):699–708

23. Saaed SA (1990) Characteristics of gypseous soil. Almuhandisson J (3):23–35

24. Seleam SNM (1988) Geotechnical characteristics of a gypseous sandy soil including the effect of contamination with some oil products. M.Sc. Thesis, Building and Construction Engineering Department, University of Technology, Baghdad, Iraq

Chapter 10

COMPRESSION AND CONSOLIDATION ANISOTROPY OF SOME SOFT SOILS

Brendan C. O'kelly

Department of Civil, Structural and Environmental Engineering, University of Dublin, Trinity College, Dublin, 2, Ireland

ABSTRACT

The compression and consolidation anisotropy of 11 soft soils were studied by conducting oedometer tests on sets of duplicate undisturbed specimens prepared in the vertical and horizontal directions from adjacent sections of carefully sampled borehole cores. The onedimensional compression, yield and creep characteristics of the various silts, clays and amorphous peaty material tested were similar for the vertical and horizontal directions. The exception was the structured, coarse fibrous peaty material which was strongly cross-anisotropic. Drainage occurred more rapidly in the horizontal direction with horizontal-to-vertical permeability ratios r_k of 1.0–1.7. Higher r_k values were associated with more marked fabrics, namely for clays with fine sand partings, fibrous organic inclusions or fine root-holes and the laminated silts. The r_k value was for practical purposes independent of the stress level.

INTRODUCTION

Soft soil deposits are often cross-anisotropic in their mechanical and drainage properties due to the preferred horizontal alignment of the solid particles during deposition and subsequent consolidation under the overburden weight. Hence, the rate of consolidation is often greater for horizontal than for vertical drainage conditions and the state of anisotropy can be assessed in terms of the horizontal-to-vertical permeability ratio r_k. Table 1 lists typical r_k values reported for soft natural clayey deposits. Accurate predictions of the amount and in particular the rate of settlement of the ground under an applied load are necessary since geotechnical design is largely driven by serviceability limit state conditions. The consolidation properties of the ground are often determined in practice using standard oedometer tests for vertical drainage conditions only. The measurement of the consolidation properties under horizontal drainage conditions receives less attention and an assessment of the state of anisotropy in terms of the permeability ratio is often made on the basis of the soil description and engineering judgment. Consequently, design predictions for the field consolidation rate are generally conservative (Cortellazzo and Simonini, 2001).

Table 1: Horizontal-to-vertical permeability ratios r_k

Description	r_k	Reference
Soft marine clay	1.05	Subbaraju (1973)
Plastic marine clay	1.2	Lumb and Holt (1968)
Bäckebol soft marine clay	1.18–1.33	Leroueil et al. (1990)
Louiseville soft marine clay	1.35	Leroueil et al. (1990)
Organic silt with peaty	1.2–1.7	Tsien (1955)
Soft clay	1.5	Basett and Brodie (1961)
Bothkennar soft silty clay	1.5–2.0	Leroueil et al. (1992)
Po soft clay	1.4–2.5	Cortellazzo and Simonini (2001)
Soft Bangkok clay	1.3–2.8	Seah and Koslanant (2003)
Singapore marine clay	2.0–3.0	Chu et al. (2002)
Po soft silty clay	2.7–4.0	Cortellazzo and Simonini (2001)

It was in this context that the degree of compression, consolidation and permeability anisotropy of various soft soils were studied by comparing the responses of duplicate sets of specimens under one-dimensional loading for vertical and horizontal drainage conditions

Test Programme

Standard oedometer tests were conducted on duplicate sets of undisturbed specimens, 76.2 mm in diameter by 19.0 mm in height, that were prepared from adjacent sections of carefully sampled borehole cores, one set carved out and tested in the vertical direction and the other set carved out and tested in the horizontal direction. The cores were recovered in 100 mm diameter aluminium tubes (12 degree cutting edge and an area ratio of 7%) using a fixed-piston sampler. The oedometer tests comprised five maintained-load stages, each stage of 24 h duration. The applied axial stress range was 12.5–200 kPa and a stress increment ratio of unity was used throughout. Specimen drainage was facilitated via porous top and bottom loading platens and the axial deformation was recorded using a displacement transducer. Table 2 lists the description, index and in situ properties of the test soils that were obtained from four sites in Ireland (Figure 1). The Carrick on Shannon site was an alluvial deposit; the Carrickmacross site a lacustrine deposit; the Shannon site an alluvial and marshland deposit; and the Waterford site an estuarine deposit. The various soils were identified by labels [1] to [11]. The small-scale fabrics of the undisturbed cores were noted following a careful visual inspection. The two peaty materials [10] and [11] were classified as H3 (spongy fibrous plant material that was only slightly biodegraded) and H6 (amorphous material that was strongly biodegraded) according to the von Post system (Head, 1992).

Table 2: Descriptions and in situ properties of test soils

Soil no.	Depth (m)	Soil description and classification (BS5930, 1999)	w_l (%)	w_p (%)	I_p (%)	G_s	w_o (%)	e_o	γ	γ_d (kN/m³)
SILT										
[1]	2.5	Very soft, laminated, calcareous SILT (marl) (MH)	68	52	16	2.31	85	2.07	14.1	7.4
[2]	4.0	Very soft, mottled grey-brown, thinly laminated, clayey SILT (MH)	64	44	20	2.67	40	1.53	14.3	10.4
[3]	2.0	Very soft, mottled grey-brown, thinly laminated SILT with some shell fragments (MH)	58	31	27	2.56	52	1.28	16.8	10.8
[4]	1.2	Very soft, grey-brown, thinly laminated SILT with some wood fibres (MV)	74	37	37	2.43	66	1.51	15.7	9.6
[5]	2.5	Soft, dark grey SILT with occassional root fibres (MEO)	91	42	49	2.45	72	1.83	14.6	8.5
[6]	3.5	Soft to firm, black, peaty SILT (MEO)	170	85	85	2.26	143	3.24	12.7	5.2
CLAY										
[7]	2.5	Very soft, grey-brown, medium laminated CLAY (CI)	42	18	24	2.72	45	1.27	17.3	11.7
[8]	5.0	Soft to firm, brown, thinly laminated CLAY with fine sand partings (CVS)	90	35	55	2.70	123	3.93	12.0	5.4
[9]	3.5	Soft, dark grey, fibrous organic CLAY (CEO)	143	52	91	2.51	134	3.55	13.1	5.4
PEAT										
[10]	1.2	Soft, coarse fibrous PEAT (von Post H3)	470	280	190	1.53	554	9.5	9.5	1.5
[11]	1.5	Very soft, fine fibrous PEAT (von Post H6)	710	380	330	1.41	712	10.3	9.5	1.2

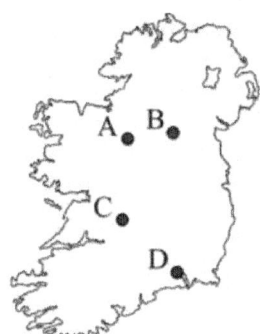

Test soil number

A : Carrick on Shannon   [5, 8, 9, 10, 11]
B : Carrickmacross   [1, 7]
C : Shannon   [6]
D : St. John's Estuary, Waterford   [2, 3, 4]

Figure 1: Locations of soft soil deposits in Ireland.

EXPERIMENTAL RESULTS

Similar initial water content, void ratio and bulk unit weight values measured for the specimen sets confirmed that they were duplicate sets. Figure 2 shows the oedometer data for the various soils. The initial specimen compression recorded for the different oedometer load stages was negligible indicating that the specimens were fully saturated. Figure 3 shows the compressibility of the soils in terms of void ratio–logarithm effective stress (e–logr') plots. Table 3 lists the compression properties for the vertical direction in terms of the primary compression index C_c, and the primary compression ratio C_c^*, (Equation 1).

$$C_c^* = \frac{C_c}{1 + e_o}$$

(1)

where e_o =the in situ void ratio and C_c =the primary compression index. The yield stress, in situ vertical effective stress (σ'_{vo}) and apparent overconsolidation ratio (OCR) are also listed in Table 3. The yield stresses for the vertical (σ'_{vc}) and horizontal (σ'_{hc}) directions were determined using the construction of compression curves after Casagrande (1936). The in situ vertical effective stress due to the overburden was calculated using the bulk unit weights and ground water level reported in the borehole log. σ'_{vo} was zero for the peaty materials [10, 11] since their buoyant weight were zero. The apparent OCR was calculated as the σ'_{vc} to σ'_{vo} ratio, a value of unity indicative of recent, normally consolidated deposits.

EXPERIMENTAL ANALYSIS

Compression And Yield Behaviour

Although the soils were marginally more compressible in the vertical direction than in the horizontal direction under one-dimensional loading, for practical purposes the responses were similar when assessed in terms of cumulative strains (Figure 2), reductions in void ratio (Figure 3) or the ratios of the horizontal-to-vertical primary compression indices (Table 3).

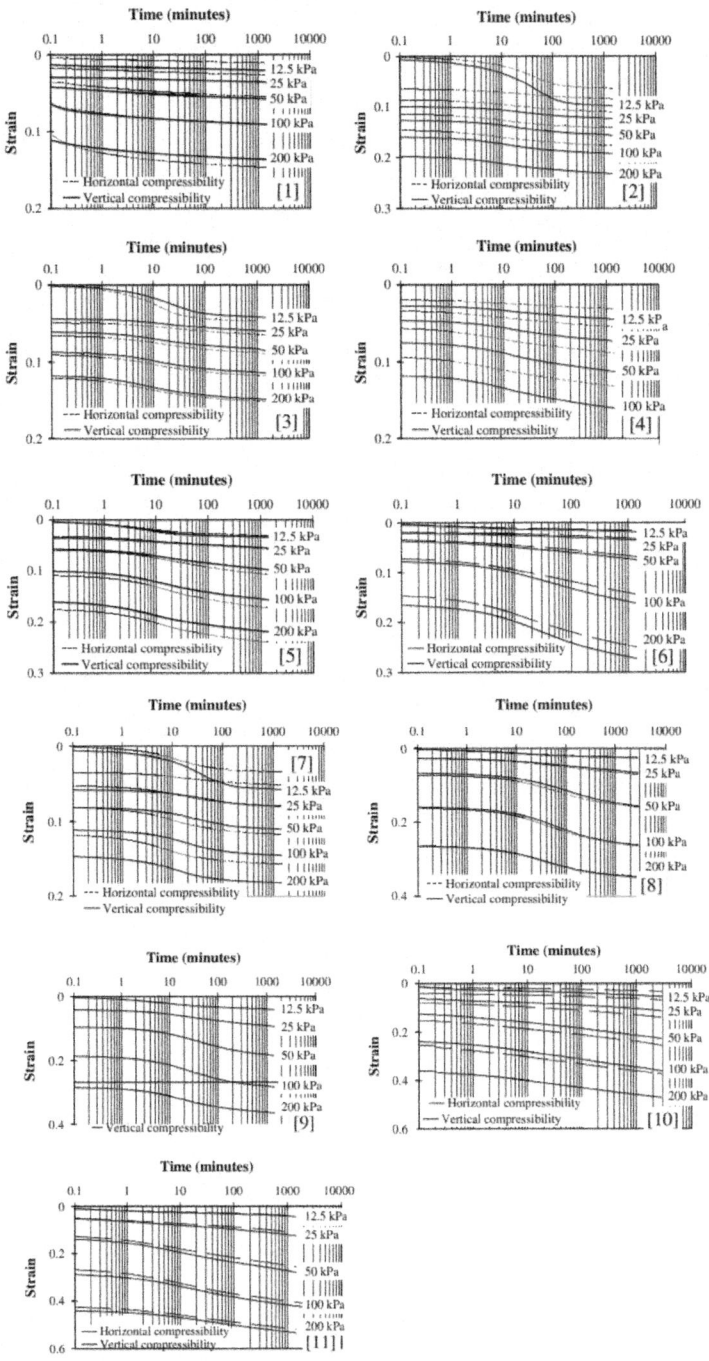

Figure 2: Cumulative strain – log time plots for test soils [1–11]

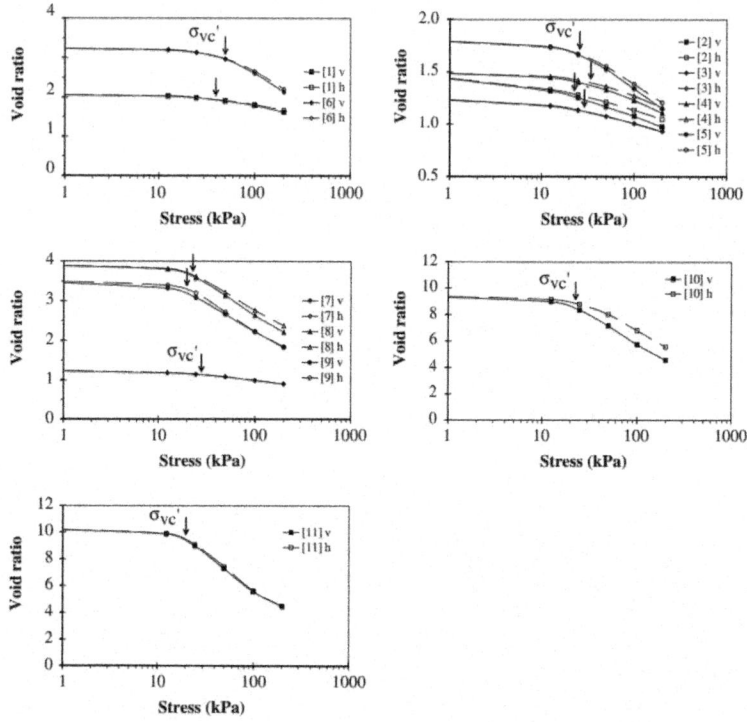

Figure 3: Compression properties for the vertical v, and horizontal h, directions.

Table 3: Compression properties for the vertical v, and horizontal h, directions

Test soil no.	Compression index		Compression ratio C_c^*	Preconsolidation pressures (kPa)		σ'_{vo} (kPa)	OCR$_v$
	C_c	$h{:}v$ ratio		σ'_{vc}	σ'_{hc}		
[1]	0.59	0.8	0.19	40	40	12	3.4
[2]	0.35	0.9	0.14	21	25	20	1.0
[3]	0.28	0.9	0.12	29	29	16	1.9
[4]	0.42	0.9	0.17	33	36	9	3.8
[5]	0.64	0.9	0.22	25	27	13	1.9
[6]	1.6	1.0	0.37	50	51	10	5.0
[7]	0.30	1.0	0.13	29	28	20	1.4
[8]	1.7	0.9	0.34	21	22	12	1.8
[9]	1.5	1.1	0.33	20	21	12	1.7
[10]	4.7	0.9	0.45	21	32	0	–
[11]	6.0	1.0	0.53	20	22	0	–

The exception was the coarse fibrous peaty material [10], which was strongly cross-anisotropic, particularly at stresses below the yield stress. In general, the yield stresses for the vertical and horizontal directions were equal since the specimen sets had experienced the same mean in situ effective confining stress. The OCR values were generally slightly greater than unity due to geological ageing. Figure 4 shows the primary compression ratioC_c^* plotted againstlogarithm c in situwater content. The data points generally located within 15% of the empirical correlation line (after Lambe and Whitman, 1979). The cores of the peaty materials [10, 11] were found to have been preloaded during piston sampling due to the cutting resistance of the fibrous plantmaterial even though the sampling tube had a sharp 12 degree cutting edge.

Primary Consolidation And Permeability

Table 4 lists the coefficient of consolidation values for the vertical direction c_{vv}, and the ratios of the horizontal-to-vertical coefficients of consolidation. The coefficient of consolidation values were calculated from interpretation of the strain–time plots using Terzaghi's one-dimensional consolidation theory. The Logarithm-of-Time curve fitting method (Casagrande and Fadum, 1940) was found to be more reliable in interpreting the data and was used instead of the Square-Root-Time curve fitting method (Taylor, 1942). Su's Maximum-Slope curve fitting method (Das, 1997) in which the data was also analysed in terms of a logarithm timescale was used for the more organic soils [6, 10, 11]. The experimental curves for these soils did not exhibit the characteristic S-shape form of the theoretical curves given by Terzaghi's consolidation, theory hence their consolidation properties could not be determined from the available data. An inspection of the compression curves corresponding to the first load stage in Figure 2 also suggested that the laminated silts [2, 3] and laminated clay [7] suffered some sampling disturbance, most likely due to the cores swelling on recovery from the ground.

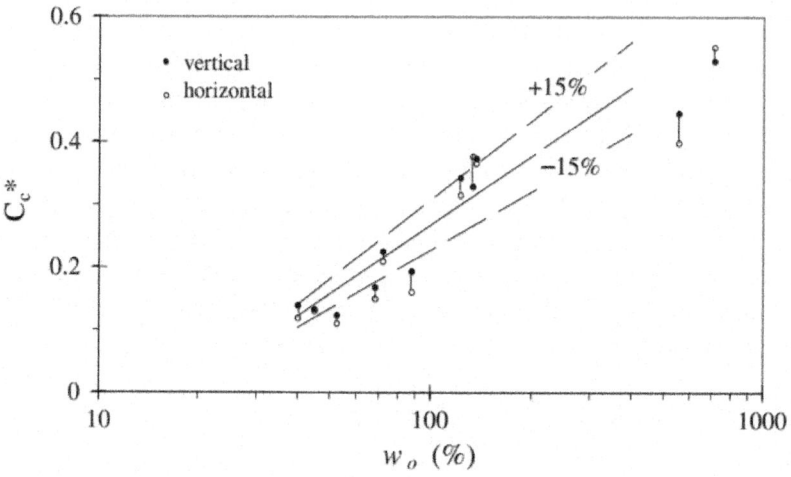

Figure 4: Primary compression ratio – log in situ water content

Table 4: Vertical coefficient of consolidation cvv, and state of anisotropy

Test soil no.	c_{vv} (m²/yr)	h:v ratio	c_{vv} (m²/yr)	h:v ratio	c_{vv} (m²/yr)	h:v ratio	c_{vv} (m²/yr)	h:v ratio	c_{vv} (m²/yr)	h:v ratio	Mean h:v ratio
[1]	370	1.0	180	1.0	90	1.0	90	1.0	30	1.0	1.0
[2]	0.4	1.1	0.5	1.2	0.5	1.3	0.7	1.3	0.8	1.3	1.2
[3]	0.8	1.0	0.8	0.9	1.0	0.9	1.0	1.0	1.2	1.1	1.0
[4]	1.3	2.4	1.0	2.0	0.8	1.9	1.0	1.9	0.9	1.7	2.0
[5]	2.9	1.0	2.9	1.0	0.7	1.2	0.7	1.1	0.3	1.1	1.1
[7]	0.4	2.0	0.3	2.0	0.4	1.9	0.4	1.7	0.2	1.5	1.8
[8]	1.5	2.0	0.6	1.7	0.1	1.6	0.1	1.6	0.1	1.5	1.5
[9]	2.0	2.3	0.5	1.9	0.3	1.2	0.3	1.2	0.1	1.2	1.6
Stress range	0 → 12.5 kPa		12.5 → 25 kPa		25 → 50 kPa		50 → 100 kPa		100 → 200 kPa		

Hence, the corresponding values of the consolidation parameters were treated with caution. Table 5 lists the coefficient of permeability values for vertical drainage conditions k_v, and the horizontal-to-vertical permeability ratio r_k. The permeability values were calculated indirectly using Equation (2) and corresponded to an ambient laboratory temperature of 21 C.

$$k_v = m_{vv}c_{vv}\gamma_w$$

(2)

where m_{vv} = the coefficient of volume change for the vertical direction; c_{vv} = the coefficient of consolidation for the vertical direction and c_w =the unit weight of water. The data in Tables 4 and 5 indicate that drainage occurred more rapidly in the horizontal direction than in the vertical direction. The exceptions were the calcareous silt [1] and shelly silt [3] which had similar permeability values for the vertical and horizontal directions. The range of mean r_k values of 1.0–1.7 for the various soils was consistent with the values reported in the literature

(Table 1). Higher r_k values of 1.4–1.7 were associated with more marked fabrics, in particular those including fine sand partings, fibrous organic inclusions or fine root-holes, namely the laminated silt [4], laminated clays [7, 8] and fibrous organic clay [9]. The r_k ratio was for practical purposes independent of the applied stress. Figure 5 shows log k_v plotted against log σ'_v (after Lambe and Whitman, 1979) which suggests an inverse log–log relationship. The lines included in Figure 5 are least-square best-fit regression lines. The significant reductions in the permeability at higher effective stresses for soils [5, 8, 9] can be explained by the closure of the fine root holes and sand-filled fissures that had facilitated preferential drainage at lower effective stresses. The data in Tables 1 and 5 can be applied in practice to obtain more accurate predictions for the rate of consolidation settlement in the field. The permeability properties measured using the oedometer apparatus for drainage in the vertical direction can be scaled by an appropriate r_k value to determine the corresponding properties for drainage in the horizontal direction. The appropriate r_k value can be estimated from Tables 1 and 5 taking into consideration the soil description, Atterberg limits and more importantly the soil fabric, noted from a careful inspection of the borehole core.

SECONDARY COMPRESSION

Table 6 lists the values of the secondary compression index C_{ae}, for the vertical direction and the horizontal-to-vertical C_{ae} ratios. The C_{ae} values were calculated as the change in the void ratio that occurred over one log cycle of time during the secondary compression phase. Definitive conclusions regarding the creep properties would have required load stages in excess of 48 h duration whereas the oedometer load stages in this study were each of 24 h duration. Nevertheless, the C_{ae} values recorded for the vertical and horizontal directions were similar for the same applied effective stress. The exception was the coarse fibrous peaty material [10] that had an initial horizontal-to-vertical C_{ae} ratio of 0.6, although this material became more isotropic at higher effective stresses due to the development of a new stress-induced fabric.

Table 5: Coefficient of vertical permeability kv, and state of permeability anisotropy

Test soil no.	k_v (m/s)	r_k	k_v (m/s)	r_k	k_v (m/s)	r_k	k_v (m/s)	r_k	k_v (m/s)	r_k	Mean r_k ratio
[1]	2E-07	0.5	7E-08	1.0	2E-08	1.3	2E-08	1.1	5E-09	1.2	1.2
[2]	9E-10	0.7	3E-10	0.9	2E-10	1.0	2E-10	1.0	1E-10	1.1	1.0
[3]	9E-10	1.2	4E-10	0.9	3E-10	0.9	2E-10	1.0	1E-10	1.1	1.0
[4]	9E-10	1.6	5E-10	1.5	3E-10	1.5	3E-10	1.6	1E-10	1.5	1.5
[5]	2E-09	1.1	2E-09	1.2	4E-10	1.0	3E-10	1.0	6E-11	1.0	1.1
[7]	6E-10	1.2	2E-10	1.5	2E-10	1.6	1E-10	1.8	3E-11	1.5	1.6
[8]	1E-09	2.0	6E-10	1.9	2E-10	1.5	1E-10	1.5	2E-11	1.0	1.7
[9]	2E-09	1.4	7E-10	1.9	4E-10	1.3	2E-10	1.3	4E-11	1.2	1.4
Stress range	0 → 12.5 kPa		12.5 → 25 kPa		25 → 50 kPa		50 → 100 kPa		100 → 200 kPa		

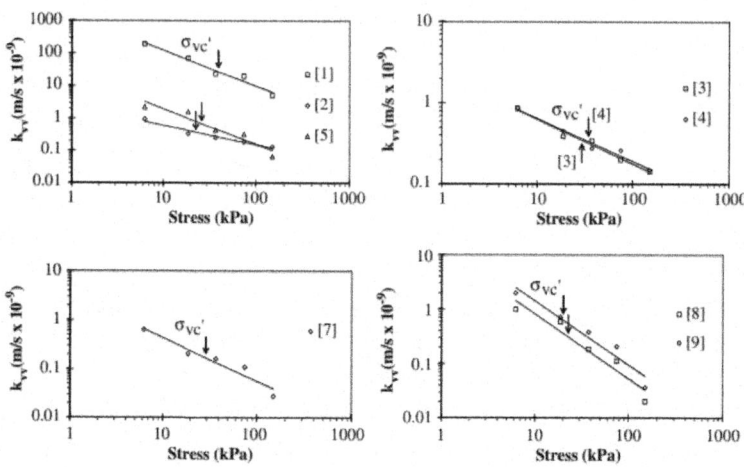

Figure 5: Coefficient of vertical permeability – log vertical effective stress.

The compressibility of the more amorphous peaty material [11] with its finer, more strongly degraded fibres was largely isotropic.

Figure 6 shows C_{ae} plotted against logarithm σ'_v for the test soils. A step increase in the C_{ae} values occurred at the yield stresses Table 6 also lists the mean values of the C_ae/C_c ratio calculated for applied stresses greater than the yield stress. The C_{ae}/C_c values agreed with the extensive work carried out by Mesri and co-workers, for example Mesri et al. (1995), with $C_{ae}/C_c = 0.02$ for the calcareous silt [1], 0.03–0.05 for the inorganic clays and silts, and 0.05–0.06 for the more organic soils [6, 10, 11].

SUMMARY AND CONCLUSIONS

The one-dimensional compression and consolidation properties of 11 soft soils from Ireland were studied by conducting oedometer tests on sets of duplicate undisturbed specimens prepared in the vertical and horizontal directions from adjacent sections of carefully sampled borehole cores.

- The compression, creep and yield properties of the various silts, clays and amorphous peaty material tested were similar since the specimens sets had experienced the same mean in situ effective confining stress. The exception was the structured, coarse fibrous peaty material which was strongly cross-anisotropic.

- Drainage occurred more rapidly in the horizontal direction with horizontalto-vertical permeability ratios r_k, of 1.0–1.7. Higher r_k values of 1.4–1.7 were associated with more marked fabrics, in particular clays with fine sand partings, fibrous organic inclusions or fine root holes and the laminated silts.

Table 6: Secondary compression index Cae, for vertical direction and state of anisotropy

Test Soil no.	C_{ae}	h:v ratio	C_{ae}	h:v ratio	C_{ae}	h:v ratio	C_{ae}	h:v ratio	C_{ae}	h:v ratio	Mean h:v ratio	Mean C_{ae}: C_e ratio
[1]	0.005	0.9	0.006	1.1	0.008	1.0	0.009	0.9	0.010	1.1	1.0	0.02
[2]	0.012	0.8	0.011	0.9	0.010	1.0	0.011	0.9	0.013	1.1	0.9	0.03
[3]	0.006	1.2	0.008	1.0	0.010	1.0	0.009	1.0	0.009	1.2	1.1	0.03
[4]	0.007	0.9	0.010	0.8	0.013	0.8	0.020	1.0	0.022	0.8	0.9	0.04
[5]	0.007	1.1	0.029	1.1	0.029	1.1	0.031	1.1	0.027	1.2	1.1	0.05
[6]	0.011	0.9	0.020	1.0	0.051	1.0	0.121	1.2	0.103	1.0	1.0	0.06
[7]	0.006	0.9	0.007	0.9	0.007	0.9	0.007	1.0	0.006	1.1	0.9	0.02
[8]	0.016	1.3	0.061	1.2	0.088	1.0	0.083	1.0	0.062	0.9	1.1	0.04
[9]	0.023	0.8	0.050	1.0	0.074	1.2	0.067	1.1	0.054	1.2	1.1	0.04
[10]	0.058	0.6	0.096	0.6	0.254	0.6	0.336	0.8	0.309	1.0	0.7	0.06
[11]	0.068	1.0	0.267	1.0	0.412	0.9	0.332	1.0	0.304	1.1	1.0	0.05
Stress range	0 → 12.5 kPa		12.5 → 25 kPa		25 → 50 kPa		50 → 100 kPa		100 → 200 kPa			

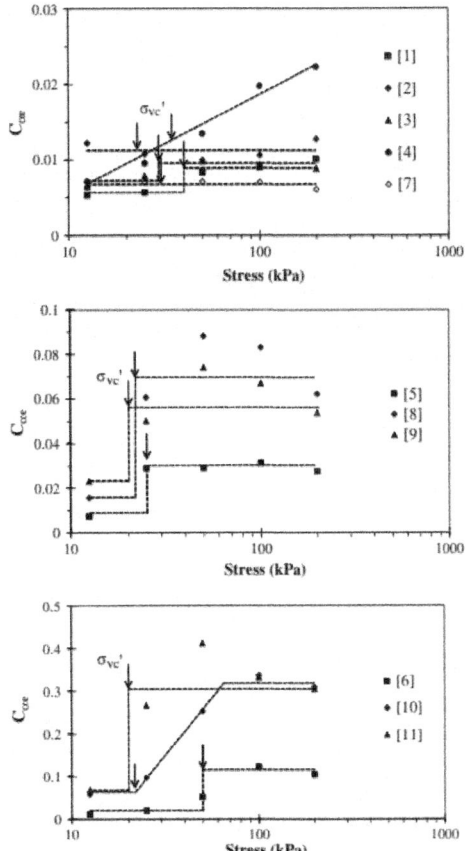

Figure 6: Secondary compression index – log vertical effective stress.

Compression of these features at higher effective stresses caused significant reductions in the permeability although the r_k values remained independent of the stress level for practical purposes.

- Preloading of the cores of peaty materials occurred during piston sampling due to the cutting resistance of the fibrous plant material

ACKNOWLEDGEMENTS

The laboratory tests were carried out by Martin Carney, Assedik Assghair and Darren Gavin and this work is gratefully acknowledged. The author would also like to thank Dr. Eric Farrell for his helpful comments during the preparation of this paper.

REFERENCES

1.   Basett, D.J., Brodie, A.F. 1961A study of Matabitchual varved clayOntario Hydro Research News1316

2.   BS5930 (1999) Code of Practice for Site Investigations, British Standards Institution, London

3.   Casagrande, A. (1936) The determination of the preconsolidation load and its practical significance, in Proceedings of the First International Conference on Soil Mechanics and Foundation Engineering, p. 60

4.   Casagrande, A. and Fadum, R.E. (1940) Notes on soil testing for engineering purposes. Harvard University Graduate School of Engineering, Publication 8

5.   Chu, J., Bo, M.W., Chang, M.F., Choa, V. 2002Consolidation and permeability properties of Singapore marine clayGeotechnical and Geoenvironmental Engineering128724732

6.   Cortellazzo, G., Simonini, P. 2001Permeability evaluation and its implications for consolidation analysis of an Italian soft clay depositCanadian Geotechnical Journal3811661176

7.   Das, B.M. 1997Advanced Soil MechanicsTaylor and FrancisWashington

8.   Head, K.H. 1992Manual of Soil Laboratory Testing: (1) Soil Classification and Compaction TestsPentech PressLondon

9.   Lambe, T.W., Whitman, R.V. 1979Soil MechanicsJohn Wiley and SonsSingapore

10.   Leroueil, S., Bouclin, G., Tavenas, F., Bergeron, L., LaRochelle, P. 1990Permeability-anisotropy of natural clays as a function of strainCanadian Geotechnical Journal27568579

11. Leroueil, S., Lerat, P., Hight, D.W., Powell, J.J.M. 1992Hydraulic conductivity of a recent estuarine silty clay at BothkennarGéotechnique42275288

12. Lumb, P., Holt, J.K. 1968The undrained shear strength of a soft marine clay from Hong KongGéotechnique182536

13. Mesri, G., Shahien, M. and Feng, T.W. (1995) Compressibility parameters during primary consolidation, in Proceedings of the International Symposium on Compression and Consolidation of Clayey Soils, Hiroshima, Vol. 2, pp. 1021–1037

14. Seah, T.H., Koslanant, S. 2003Anisotropic consolidation behavior of soft Bangkok clayGeotechnical Testing Journal26266276

15. Subbaraju, B.H. (1973) Field performance of drain wells designed expressly for strength gain in soft marine clays, in Proceedings of the Eight International Conference on Soil Mechanics and Foundation Engineering, Vol. 2.2, pp. 217–220

16. Taylor, D.W. (1942) Research on consolidation clays. Department of Civil and Sanitation Engineering, Massachusetts Institute of Technology, Report 82

17. Tsien, S.I. 1955Stability of marsh depositsHighway Research Board, Bulletin151543

Chapter 11

GEOTECHNICAL, CHEMICAL AND MINERALOGICAL EVALUATION OF LATERITIC SOILS IN HUMID TROPICAL AREA (MFOU, CENTRAL-CAMEROON): IMPLICATIONS FOR ROAD CONSTRUCTION

Brice T Kamtchueng[1],[2], Vincent L Onana[2] , Wilson Y Fantong[3] , Akira Ueda[1] , Roger FD Ntouala[2] , Michel HD Wongolo[2] , Ghislain B Ndongo[2] , Arnaud Ngo'o Ze[2] , Véronique KB Kamgang[4] and Joseph M Ondoa[2]

[1] Department of Earth, Life and Environmental Sciences, Graduate School of Science and Engineering for Education, University of Toyama, Gofuku 3190, Toyama 930-8555, Japan

[2] Department of Earth and Life Sciences, Faculty of Sciences, University of Yaounde I, PO Box 812, Yaounde, Cameroon

[3] Institute of Mining and Geological Researches (IRGM), PO Box 4110, Yaounde, Cameroun

[4] Higher Teacher's Training College (ENS), University of Yaounde I, PO Box 4110, Yaounde, Cameroun

ABSTRACT

Increased cost associated with the used of high quality materials have led to the need for local soils to be used in civil engineering works. In this paper, geo-chemical approaches coupled with conventional geotechnical techniques has been used to investigate vertical and lateral peculiar engineering index properties (EIP) of micaschist derived soils from Mfou, Central-Cameroon for their uses in road construction. The X-ray diffraction (XRD) analysis conducted on the soils indicated the absence of swelling clays. The main mineral phases were quartz, kaolinite, goethite, magnetite and chlorite. Geo-chemical results show that the investigated soils are "true laterites" made up of 41 wt.% of Fe_2O_3, 35 wt.% of SiO_2, 21 wt.% of Al_2O_3, 1.17 wt.% of K_2O and 0.05 wt.% of CaO. The results of geotechnical tests suggest that the upper clayey layer (UCL) and bottom mottled clayey layer (MCL) of the weathering profiles are poorly graded soils with EIP (fines particles (FP) of 61 and 63%, plasticity index (PI) of 30 and 31%, Californian Bearing Ratio (CBR) at 95% of 21 and 19%), which do not allow their use as raw materials in road construction whereas, intermediate nodular layer (INL) are well graded soils having EIP (FP of 26%, PI of 26% and CBR of 39%) that meet the specification required for sub-base materials for light traffic roads. The relatively high sesquioxyde present in these residual soils may act as cementing agent, thereby making the compacted soils relatively brittle. The direct shear test results show that the soils have high bearing capacity (cohesion of 62 Kpa and 27.2° angle of internal friction) making them to be useful in slope stability and shallow foundation design. The comparison of the studied soils with some lateritic soils in Sub-Saharan Africa indicates that i) genesis and climatic conditions are potentials factors that influence EIP of lateritic soils, ii) lateritic soils developed under semi-arid conditions exhibits EIP better than those developed under tropicaland sub-tropical conditions.

INTRODUCTION

A considerable increase in soil utility for engineering works is expected as any country aspires towards improved infrastructural development. The relationship between all engineering infrastructure and their foundation soils is of paramount importance for designers and contractors. Incessant occurrence of road pavement deterioration and building collapse, mainly because of their poor geotechnical and mechanical properties has made it imperative for a proper understanding of the geotechnical properties of soils (Garg2009). Soil is defined as a three-dimensional body with properties that reflect the impact of mankind, climate, vegetation, fauna and relief on the soil's parent material over a variable time span (Deckers et al. 2001). The nature and relative importance of each of those "soil forming factors" varies in time and space. As far back as the eighteenth century, geologists identified that in warm, moist, temperate and tropical climates water percolating through rock has a strong weathering action (Millard 1993). Chemical reactions increase with an increase in rainfall and temperature, and accordingly soils from the tropics exhibit different engineering properties (Millard 1993). Despite the great effort that has been made by previous works to classify and differentiate tropical soils, a uniform nomenclature /classification system does not exist yet for these soils. Hence, the term "lateritic clays", "lateritic gravels" and even "laterites" are still used by engineers to describe any reddish tropical soils (Northmore et al. 1992). Fortunately, for the engineering purposes it does not matter whether the classification is correct, but that the geological and engineering properties as predicted or derived from testing are reliable.

The soils used in road construction in "Sub-Saharan Africa" are mostly lateritic soils including lateritic gravels, lateritic clays and lateritic shales. Lateritic soils contribute to the general economy of the region where they are found (Lemougna et al. 2011). Their scope is very wide and include civil engineering, agronomic, mining research (iron, aluminum and manganese) deposits. In a tropical area such as Cameroon that is in a phase of infrastructural development, lateritic soils has been, and still a topic of interest and discussion. Due to the natural relative abundance of these soils, availability, favorable engineering properties, they have been very useful for construction of foundation, roads, airfields, low-cost housing and compacted fill in earth embankments (Gidigasu 1983; Autret 1983; Ekodeck 1984; Sikali and Mir-Emarati1986; Bagarre 1990; Tardy 1992 Tockol et al. 1994; Ojo and Adeyemi 2003).

The road network in Cameroon covers a distance of about 50,000 km. Which About 5,000 km is paved, 18,016 km is unpaved, and 27,693 km are rural roads (Transport Policies in Sub-Saharan Africa" (TPSSA),2004). According to the "TPSSA (2004)» report, about 90% of the total road network in Cameroon is made up of unpaved roads. Thus this limits trade exchanges between border countries and the link between the hinterland (high agricultural productivity area) and the biggest cities. Consequently, constraint the socio-economic development of the country. While several research activities have been done with regards to the engineering index properties (EIP), reliability and durability of soils used in road construction throughout the world (eg. Autret 1983; Bagarre 1990; Tardy 1992; Ojo and Adeyemi 2003; Bayewu et al. 2012), only few studies revealed the geotechnical characteristics of lateritic soils in Cameroon (eg. Sikali and Mir-Emarati1986; Ekodeck 1985; Mbumbia et al. 2000; Onana 2010; Lemougna et al. 2011; Logmo et al. 2013). The results of their investigations showed that EIP of the lateritic soils in Cameroon vary according to the nature of the parent rocks and local climatic regime. Generally, they are of good to fair quality as raw materials used both in building and road constructions. However, as observed by Millogo et al. (2008) and references therein, lateritic soils chosen based only from the results of geotechnical tests behaved badly on the pavements after a short period. This behavior could be attributed to the insufficiency of the geotechnical tests justifying the choice of lateritic soils destined to the road construction. To backup this short coming, some authors (eg. Tockol 1994; Mahalinger-Iyer and Williams 1997; Millogo et al. 2008) suggest that chemical and mineralogical analysis must be conducted on lateritic soils in addition to the geotechnical classic tests. Accordingly, the purpose of this work is to combine geotechnical tests to chemical and mineralogical analysis to characterize micashist derived lateritic soils from Mfou, Central-Cameroon. Data from this study will not only contribute to evaluate the suitability of these soils for road construction but it will also provide baseline data for engineers, designers and contractors since the work was carried out when the road construction section Mfou-Nsimalen was on-going.

Study area

Administratively, Mfou is located in the Mefou and Afamba Division in the South-Eastern part of Yaounde. Geographically, the area under consideration lies between latitudes 3°40' and 3°45' and longitudes 11° 35' and 11°40' E, covering an estimated area of 85 km² (Figure 1).

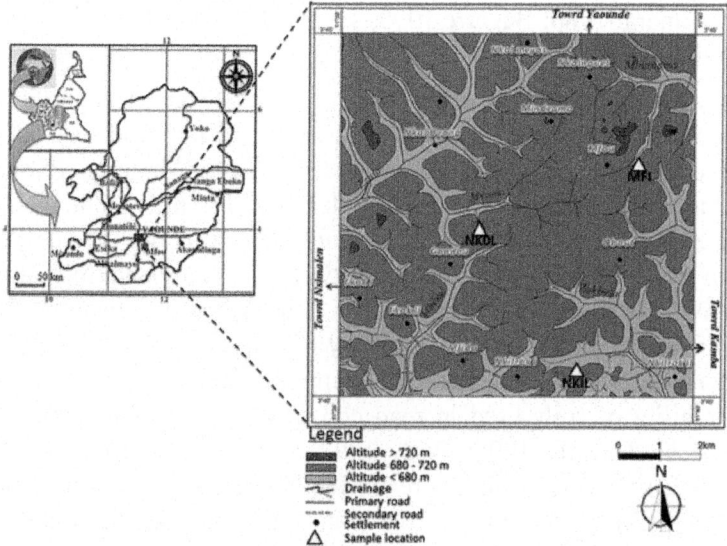

Figure 1: Location and topographic map of the study area showing sampling sites.

Despite being located at about 300 km from the coast, the area is subjected to maritime influence (Ndam et al. 1998). The climatic regime is Guinea equatorial and bimodal. It is characterized by four distinct seasons including two rainy season (from August to November and from April to June) and two dry season (from December to March and from June to July). The annual rainfall, annual average temperature and annual average relative humidity are 1600 mm, 24°C and 79%, respectively (Molua and Lambi 2006). The vegetation belongs to the dense moist forest type described by Vallerie (1995) as a semi-deciduous forest with Sterculiaceaes and Ulmaceaes as dominant species. As observed by Segalen (1967), this climatic regime promotes the physico-chemical alteration of the rocks in the southern part of Cameroon. The topography of the study site varies within a narrow range. It is characterized by asymmetrical gentle slopes and dissected shallow valleys. The land use is dominated by agricultural activities. The hydrographic network is heterogeneous and dense showing dendritic, parallel and rectangular facies.

The detailed background regarding the geology of the study area can be found in previous studies (eg., Gazel and Guiraudie 1967; Segalen 1967; Gidigasu 1983; Nzenti et al. 1988; Ekodeck 1984; Yongue 1986; Kamgang 1987; Onana 2010). Briefly, the basement is made up of metamorphic rocks of the central Neoproterozoic panafrican fold belt (Nzenti et al. 1988). These metamorphic rocks have been identified as greenschist bearing micaschist (Gazel and Guiraudie 1967).

EXPERIMENTS

Sampling methodology

Prior to the sampling campaign, samples collection sites were selected on the 1:50,000 topographic map of Yaounde (sheet 4a NA – 32 – XXIV; source: National Institute of Cartography of Cameroon). Sample collection was undertaken in November 2010 where vertical and lateral variations of the weathering profiles were investigated. As shown in Figures 1, three holes (NKIL, NKOL and MFI) were dug and nine samples were collected from the upper clayey layer (UCL), intermediate nodular layer (INL) and bottom mottled clayey layer (MCL) of the weathering profiles. Undisturbed samples were collected from the UCL following the method described by Robitalle and Tremblay (1997). The method involved the placing of square template over the area to be sampled. The sample place was cleaned to remove the organo-mineral layer. A wood square template was fixed carefully on the pre-drilled hole in each corner to form a cage around the sample. The soil was excavated around the cage and the wood square template was later on carefully removed to avoid disturbance in the soil's structure. A polyvinyl chlorite (PVC) cylinder with a diameter of 4.9 inches and length of 1 inch was inserted in the obtained cone of soil and later separated from it bottom part with a side walk scraper. Both sides were sealed with paraffin to prevent the exchange of moisture content between the sampled soil and the atmosphere. Disturbed samples were collected after undisturbed samples. At each site, samples were collected on a flat terrain to minimize the influence of topography on the engineering properties of the soils. The samples were then air-dried for 3 weeks and later subjected to engineering tests. Eighty kilograms of sample was collected from each layer of the holes and stored in bags. A sample of 10 kg of fresh rock was collected for petrographic studies. The color of the soil samples was determined using a Munsell color chart (1975).

Description and method of laboratory tests

Preparation of thin sections was carried out at the Institute of Geological and Mining Research (IRGM, Cameroon). Geotechnical and mechanical tests were done at the National Laboratory of Civil Engineering (LABOGENIE, Cameroon). X-ray diffraction (XRD) and X-ray fluorescence (XRF) were done at the Laboratory of Environmental Science and Engineering of University of Toyama (Japan).

Mineralogical and chemical analyses

Mineral content is the main factor controlling the size, shape, physical and chemical properties of soil mechanics. Two thin sections were prepared for the parent rock to decipher the texture, mineralogy and alteration. The slides were closely examined under plane polarized light. The stage of the microscope was rotated continuously to attain different views of the slides. Photomicrographs from each slide were taken under cross nicols to ascertain the compositional features. Detailed identification and description of the mineral composition were determined and their percentages were estimated.

Prior to XRD and XRF analysis, the samples were dried to remove moisture content and powdered until grain pass through sieve 80 μm. The XRD analyses was done using a Bruker D8 DISCOVER-TUS diffractometer equipped with a D8 goniometer with a precision of $1/10000°$ by a stepping motor and an optical encoder. The scan performed was recorded in 2θ in a range between 5 and 70 ° for a period of around 4 hours. The diffractograms were treated with DIFFRAC plus software (version 1.2.28).

The percentage of oxides (wt.%) in the soils was determined by X-ray spectrometric powder technique described by Leake et al. (1970). The test was performed using an X-Ray Spectrometer model EDX-700HS with Energy Dispersive. Only samples from intermediate nodular layer (INC) were subjected to the XRD and XRF analyses.

Engineering tests

Engineering tests were carried out to know the suitability of the sampled soils as compacted materials and for road performance. This involves i) the densimetric tests (specific gravity), ii) classification tests (grain size distribution, plasticity) and iii) geotechnical tests (compaction test, California Bearing Ratio and direct shear test).

The specific gravity which is the measure of the density of a soil relative to that of water was estimated by repeated weighing method as described by Bouassida and Boussetta (2007). Fifteen gram of the soil samples that passed through sieve of 0.425 mm was added to the pycnometer and weighed. Sufficient air-free distilled water was added to soil sample in the bottle and shaken to eliminate air indirection. The bottle and its content was weighed. The pycnometer was later filled with distilled water and weighed. To calculate the specific gravity of the samples, the following mathematical equation was used:

$$Y_s = \frac{P_2 - P_1}{(P_4 - P_1) - (P_3 - P_2)}$$

(1)

where P_1 is the weight of the pycnometer, P_2 is the weight of pycnometer containing sample, P_3 is the weight of the bottle containing air-free distilled water and sample and P_4 is the weight of the pycnometer filled with distilled water

The grain size distribution was determined following the NF P18-560 standard (AFNOR 1978) for particle size distribution by dry sieving and the NF P94-057 standard (AFNOR 1992) for particle size distribution by sedimentation. The liquid limit (LL) was measured and the plastic limit (PL) following the NF P94-051 standards (AFNOR 1993).

As recommended by the handbook of pavement design for tropical countries (CEBTP 1984), modified Proctor test was chosen for this study. The optimum moisture content (OMC) and the maximum dry density (MDD) were determined according to NF P 94–093 standard (AFNOR 1999).

The Californian Bearing Ratio test (CBR) after 4 days soaking with a surcharge weight of 5.5 kg was determined according to NF P94-078 standard (AFNOR 1997). The punching was achieved with a press Seditech made with a manometer for pressure measurement.

Direct shear test was used to determine a quantity known as the maximum past stress, which is often assumed to be the greatest effective overburden stress to which soil has ever been subjected. This test was determined using undisturbed samples following the standard method described by the NF P94-071-2 normalization (AFNOR 1999). Only the UCL sample was subjected to this test.

RESULTS AND DISCUSSIONS

Stratification

The weathering profile is made up of four visible layers (Figure 2): i) saprolite mottled clay layer, ii) intermediate nodular layer, iii) upper soft clayey layer and iv) organo-mineral, which has not been characterized. In the saprolite mottled clay layer (MCL), the micaschist architecture is partially preserved. This layer composed of red materials (5YR 5/8) with yellow patches. Altered garnets, quartz crystals and muscovite flakes are the most visible minerals of this layer. Transition with overlying layer is gradual and wavy. The intermediate nodular layer (INL) of about 1 m thick consists of reddish (5YR 5/8) sandy clay material containing nodules embedded in a mottled matrix with various shapes. Coarse grains of centimentric size represent about 70% of the total volume of this layer. The transition with over lying layer is sharp and wavy. The upper clay layer (UCL) had thickness that vary from 0.1 to 1 m and made up of sandy clay material. This layer is a porous reddish brown soil (7.5 YR 6/6). The clay matrix of this layer is poor in quartz grains.

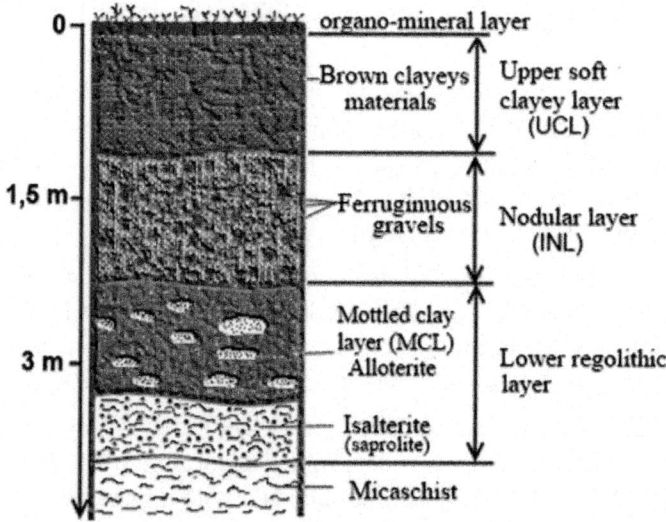

Figure 2: Schematic structure of the weathering profile of the soils in Mfou.

In general, the lateritic soils in the study area are yellowish to brown-reddish in color and contain high percentage of iron oxides as discussed later. These pigmentations are consistent with the observation of De Graff-Johnson (1972) and Tardy (1992) on the lateritic soils formed under oxidized conditions and good drainage. Their strong pigmentation may also indicate the good aggregation of the soils due to their strong surface charge properties (Tardy 1992).

Petrographic characteristics of rock in the study area

The bedrock of the study area is composed of quartz, biotite, muscovite, garnet, chlorite, sillimanite, K-feldspar (orthoclase and microcline), and plagioclase (Figure 3). Accessory minerals are apatite, zircon and sphene. The textures of the sampled micaschist are lepidoblastic, granoblastic and porphyroblastic with sizes of minerals varying from a few mm to a few cm (Figure 3). Quartz exhibits undulatory extinction and occurs as an interpenetrative mosaic of almost equal grains, up to 1 mm in size that defines ribbons parallel to foliation planes. This mineral constitutes about 35 to 40% in both the matrix and as inclusions in garnet.

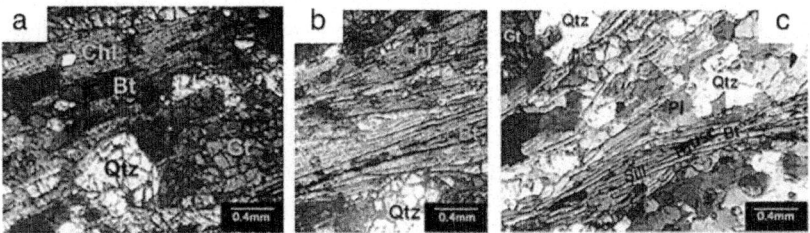

Figure 3: Photomicrographs of micashist of Mfou showing various textures. (a) lepidoblastic texture, **(b)**granolepidoblastic texture, **(c)** porphiroblastic textures. Qtz: quartz, Chl: chlorite, Gt: garnet, Music: muscovite, Bt: biotite, Sill: Sillimanite, Pl: plagioclase.

Biotite and muscovite make up 30% volume of the rock sample and vary in length from about 0.1 to 2 cm (Figure 3a and c). Two generations of micas were visible: 1) deformed flakes of up to 2 cm in size, which are most probably relics of the clasts protolith, 2) a second generation, in the matrix, is fine to medium grains. An anastomising configuration of biotite and muscovite around large garnet porphyroblasts (Figure 3a) indicates that schistosity postdated crystallization of garnet. Biotite and muscovite also occur in fractured garnets. Biotite is usually replaced with chlorite and fine grained ferrous oxides (Figure 3b).

Garnet occurs as subhedral to anhedral porphyroblasts from 0.2 to 10 mm in size, and are sometimes partially replaced by aggregates of chlorite and biotite (Figure 3a). Some garnet crystals contain sigmoidal to spiral quartz, micas and ilmenite, suggesting a synkinematic crystallization over printing crenulation cleavage (Ferry and Spear 1978). Most garnets are oriented parallel to the dominant foliation.

Mineralogical and chemical characteristics

The mineral assemblages and their modal composition are quartz (35%), muscovite (25%), biotite (15%), garnet (10%), chlorite (6%), sillimanite

(3%), plagioclase (3%) and K-feldspar. The diffractogram obtained from the XRD analysis is shown in Figure 4. The X-ray patterns showed that the soils under investigation consist of quartz, kaolinite, goethite, magnetite and chlorite. This result is in agreement with previous finding reported on lateritic soils in south Cameroon (Ekodeck 1984; Onana 2010; Lemougna et al. 2011; Tsozué et al. 2012). Tsozué et al. (2012) showed that iron oxides in lateritic soils of south Cameroon are always associated to kaolinite. These authors further explained that the presence of goethite in lateritic soils developed on micashist is the result of the disruption of kaolinite-hematite association which characterizes the primary ion duricrust under hydratation. The mineralogical investigations also indicated the absence of swelling clays such as montmorillonite, hence undesirable swelling characteristics seems not expected in these soils. Nevertheless, the presence of goethite and other iron oxides existing in these residual soils may act as cementing agents, which as a consequence may contribute to tensile cracking by making the compacted structure relatively brittle. Indraratna and Nutalaya (1991) argued that the presence of high goethite content in the residual soil derived from weathered igneous and metamorphic rocks may have influenced cracking of the Shek Pik Dam in Hong Kong, China.

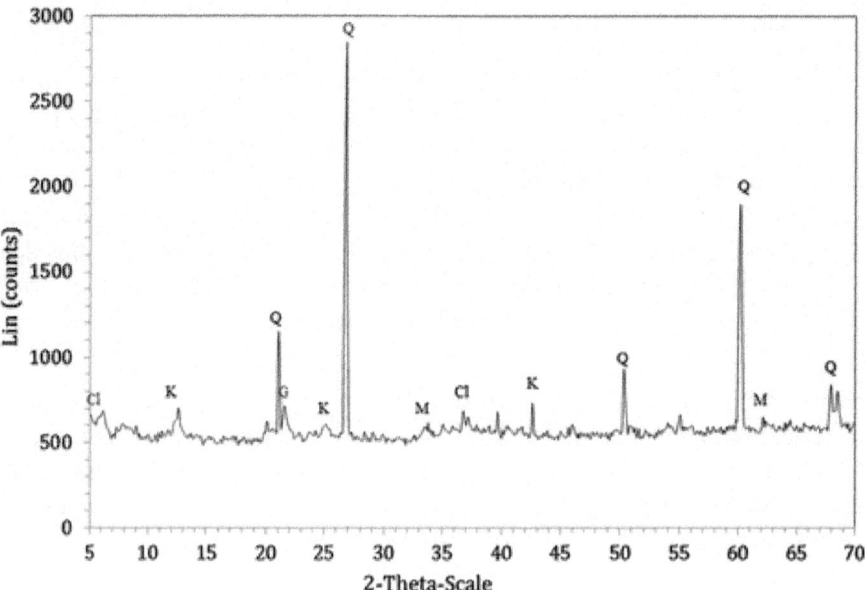

Figure 4: X-ray diffraction spectrum of < 80 μ m samples fraction of soils in Mfou. Q = quartz, M = magnetite, G = goethite, K = kaolinite, Cl = chlorite.

The chemical composition of the samples given in Table 1 shows the presence of 41 wt.% of Fe_2O_3 followed by 35 wt.% of SiO_2 and 21 wt.% of Al_2O_3 and feeble content of 1.17 wt.% of K_2O and 0.05 wt.% of CaO. This result agrees with the petrographic and mineralogical studies which indicated that quartz, iron-rich minerals and alumino-silicates are the dominant phases in both the parent rock and its derived products. Rossiter (2004) compiled the classification of soils according to the degree of laterization by evaluating silica–sesquioxide (S-S) ratio (SiO_2 / ($Fe_2O_3+Al_2O_3$)) (Table 1). Accordingly, soils having S-S ratio>2 are considered as non-lateritic soils. For lateritic soils, this S-S ratio lies between 1.33 and 2 and for true laterites the ratio is<1.33. The soils under investigation have S-S ratio (average of 0.6) lower than 1.33 suggesting that they are "true laterites" soils which has not undergone a considerable degree of laterization (Rossiter 2004).

Table 1: Chemical composition of the studied soils

Loca-tion	SiO_2	Fe_2O_3	Al_2O_3	TiO_2	K_2O	Cr_2O_3	ZrO_2	CuO	CaO	MnO	ZnO	Rh_2O_3	Total	S-S ratio
Nkil	41.72	34.28	20.55	1.67	1.29	0.16	0.09	0.07	0.07	0.04	0.03	0.01	99.99	0.76
Nkol	30.84	44.43	21.53	1.70	1.02	0.16	0.09	0.06	0.04	0.07	0.03	0.01	99.98	0.46
Mfi	32.05	43.94	20.51	1.79	1.22	0.14	0.11	0.05	0.04	0.08	0.02	0.01	99.96	0.49
Aver-age	34.87	40.88	20.86	1.50	1.17	0.15	0.09	0.06	0.05	0.06	0.02	0.01	99.72	0.57

Oxides composition in weight percent, S-S: silica-sesquioxyde.

ENGINEERING CHARACTERISTICS OF SOILS

Specific gravity

The results of the specific gravity of the investigated samples are presented in Table 2. The values range from 2.62 to 2.85, 2.34 to 2.85 and 2.07 to 2.80 for Nkil, Nkol and Mfi profiles, respectively. The lowest (average of 2.38 gf/cm³), and the highest values (average of 2.83 gf/cm³) of the specific gravity are found in UCL and INL samples, respectively. The results of studies conducted on lateritic soils (eg. Gidigasu 1983; Tockol 1994; Kabir and Taha 2004; Ramamurthy and Sitharam 2005; Que et al. 2008) have shown that specific gravity is closely linked with the mineralogy and/or the chemical composition of the soil. According to De Graff-Johnson, (1972), the higher the specific gravity, the higher the degree of laterization. Furthermore, the larger the clay fraction and the alumina concentration, the lower is the specific gravity. Thus, INL samples appear to be the most evolved samples in term of laterization (ferruginization) process.

Table 2: Specific gravity values of studied soils

	Specific gravity				
	Nkil	Nkol	Mfi	Average	SD
Upper clayey layer (UCL)	2.62	2.34	2.17	2.38	0.23
Intermediate nodular layer (INL)	2.84	2.85	2.80	2.83	0.03
Bottom clayey layer (MCL)	2.85	2.81	2.07	2.57	0.4

Nkil, Nkol and Mfi: sample location site, SD: standard deviation.

Grain size distribution and Atterberg's limits

Grain size distribution is one of the most important elements in the design of structures on, in, or composed of soils (Naresh and Nowatzki 2006). The results of grain size distribution analysis are given in Table 3. The statistical summary of obtained results are presented in Table 4 and represented graphically in Figure 5. The percentage of fines (% passing sieve 80 μm) has significant effect on the performance of the base / sub-base materials (Garg 2009). Too much fines will result in the reduction in the possible maximum density and strength and increase the susceptibility to weakening from water infiltration or seepage (Garg 2009). The results show that the amount of fines in UCL and MCL soil samples are quite similar (greater than 60%). Figure 5 shows that UCL and MCL are clayey soils that are poorly graded whereas INL are well graded gravel and sandy soils. The "well graded" curve (INL soil samples) represents a non-uniform soil with a wide range of particle sizes that are evenly distributed (Figure 5). Densification of a well-graded soil causes the smaller particles to move into the voids between the larger particles (Naresh and Nowatzki 2006). As the voids in the soil are reduced in the INL soil samples, the density and strength of the soil may increase. In contrast, poorly graded or uniform soils (UCL and MCL soil samples) are composed of a narrow range of particle sizes (Figure 5). During compaction of these soils, inadequate distribution of particle sizes prevents reduction of volume of voids with infilling by smaller particles. Thus, uniform soils are expected to have low mechanical properties. As clearly shown in Figure 5, only samples from INL meet the limit of not more than 35% of fines recommended by the CEBTP for sub-base materials (Table 5). None of the tested samples meet the CEBTP specification of not more than 20% of fines for base materials.

Table 3: Geotechnical and mechanical results of the studied soils

	Samples location	Depth of sampling (m)	Particles size distribution (%)							Atterberg's limits				Modified proctor		NMC (%)	CBR at 95% of MDD	Classification	
			12.5 (CG)	5 (FG)	2 (CS)	0.5 (MS)	0.08 (FS)	0.05 (Silt)	0.002 (Clay)	LL (%)	LP (%)	PI (%)	CI	MDD (g/cm³)	OMC (%)			AAS-HTO	GTR
UCL	Nkil	0.50-0.80	100	100	99.3	92.1	68.4	63.9	46.8	60.8	30.7	30.1	1.20	1.67	20.8	24.6	21.0	A-7-5	A₃
	Nkol	0.10-0.35	100	100	98.9	95.4	64.7	60.3	44.2	58.3	30.4	27.9	1.24	1.6	18.3	23.7	15.0	A-7-5	A₃
	Mfi	0.60-1.10	100	99.6	97.2	94.3	66.8	61.3	47.8	62.6	32.1	30.5	1.19	1.72	16.5	26.2	26.0	A-7-5	A₃
INL	Nkil	1.15-1.70	97.6	72.2	41.8	33.3	26.7	22.7	15.4	59.8	32.4	28.4	1.7	2.1	11.2	13.4	38.0	A-2-7	B₆
	Nkol	0.50-0.80	88.8	65.1	54.9	35.3	30.2	28.0	18.4	55.9	30.6	25.3	1.5	2.0	13.7	16.8	24.0	A-2-7	B₆
	Mfi	1.20-1.60	92.5	63.8	34.3	25.4	21.1	18.9	14.6	50.9	28.2	22.7	1.65	2.08	11.8	12.3	54.0	A-2-7	B₆
MCL	Nkil	2.70-3.30	100	98.8	93.6	83.6	63.6	58.6	43.5	62.4	31.8	30.5	1.3	1.82	18.9	21.9	16.0	A-7-5	A₃
	Nkol	1.90-2.30	100	97.6	94.3	81.9	69.8	65.1	47.4	61.2	33.1	28.1	1.26	1.74	19.5	25.6	17.0	A-7-5	A₃
	Mfi	2.40-3.50	98.3	94.8	90.2	84.6	61.9	60.4	44.3	66.7	33.8	32.9	1.34	1.70	15.4	22.3	23.0	A-7-5	A₃

Nkil, Nkol and Mfi: sample location site, CG: coarse gravel, FG: fine gravel, CS: coarse sand, MS: medium sand, FS: fine sand, LL: liquid limit, LP: plasticity limit, PI: plasticity index, CI: consistency index, MDD: maximum dry density, OMC: optimum moisture content, NMC: natural moisture content, AASHTO: American Association of State Highway and Transportation Officials, GTR: Guide des Terrassements Routers.

Table 4: Statistical summary results of geotechnical and mechanical tests

Geotechnical parameters	Results		
	UCL (n=3)	INL (n=3)	MCL (n=3)
Particles size distribution (%)			
< 2 mm (skeleton)	98.5	43.6	92.7
< 0,5 mm (mortar)	93.9	31.3	83.4
< 80 μm (fines particles)	66.6	26	65.1
< 50 μm (silt)	61.8	24.7	61.4
< 2 μm (clays)	46.3	16.1	45.1
Atterberg's limits			
Liquid limits, LL (%)	60.6	55.5	63.4
Plasticity limits, PL (%)	31.1	30.4	32.9
Plasticity Index, PI (%)	29.5	25.5	30.5
Consistency Index, CI	1.2	1.7	1.3
Modified Proctor			
Optimum moisture content, OMC (%)	18.5	12.2	17.9
Maximum dry density, MDD (g/cm^3)	1.7	2.1	1.8
CBR test			
CBR at 95% of MDD (%)	21	39	19
Natural moisture content, NMC (%)	24.8	14.2	23.3
Classification of soils			
Highway Research Board (AASHTO)	A-7-5	A-2-7	A-7-5
Earthworks Road Guide (GTR)	A3	B6	A3

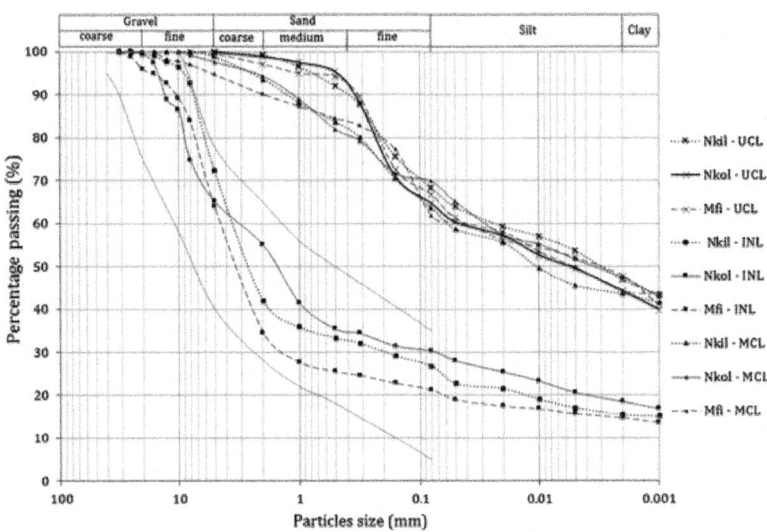

Figure 5: Sieve curves of soils samples from different layers of the weathering profiles in Mfou. UCL: upper clayey layer, INL: intermediate nodular layer, MCL: bottom mottled clayey layer. The continue gray curves represent the upper and lower bounds of typical lateritic soils intended to be used in road construction according to the road pavement standards for tropical countries (CEBTP, 1984)

Table 5: Compliance of lateritic soils derived from micaschist of Mfou to road pavement standards for tropical countries (CEBTP, 1984)

Engineering parameters	Base materials	Sub-base materials	Samples		
			UCL	INL	MCL
Atterberg's limits					
Liquid limits, LL (%)	35 max	60 max	60.6	55.5	63.4
Plasticity Index, PI (%)	15 max	30 max	29.5	25.5	30.5
Swelling potential (%)	0.3 max	1 max	2.1	1.4	2.1
Grain size distribution (%)					
20 mm	60-100	75-100	100	99	100
10 mm	35-90	58-100	100	91	100
5 mm	20-75	40-78	99	67	99
2 mm	12-50	28-65	98	43	93
0.080 mm	4-20	5-35	67	26	65
Modified Proctor					

Maximum dry density, MDD	2.00 min	1.80 min	1.7	2.1	1.8
(g/cm³)	80 min	30 min	21	39	19
CBR at 95% of MDD (%)					

Min: minimum, Max: maximum.

The Atterberg consistency limit tests show that the UCL and MCL samples have in average a LL of 61% and 63%, PL of 31% and 33%, plasticity index (PI) of 29% and 30% while INL samples has in average a LL of 55%, PL of 30% and PI of 26% (Table 4). The results of Atterberg's limit tests together with the grain size distribution results allowed the classification of UCL and MCL soil samples in the A-7-5 group of American Association of State Highway and Transportation Officials (AASHTO) classification scheme and in the A_3 group of the "Guide des Terrassements Routiers" (GTR) classification. These groups are described as fair to poor clayey and silty soils. The INL soil samples belong to the class A-2-7 and B_6, respectively according to the aforementioned classifications and are defined as gravelly and sandy soils. However, all the tested samples do not meet the CEBTP specification of not more than 15% plasticity index for base materials. Similarly, the results also indicate that the liquid limit of all the investigates samples are nearly within the range of not more than 60% for sub-base and exceed the maximum limit of 35% specified by the CEBTP for base materials (Table 5). Further, the plot in the Casagrande diagram (Figure 6) shows that all the samples are located along the A-line on the limit of high plasticity clays and silts domains. According to Day (2000) lateritic soils consisting mainly of un-swelling clay such as kaolinite are generally located below the A-line.

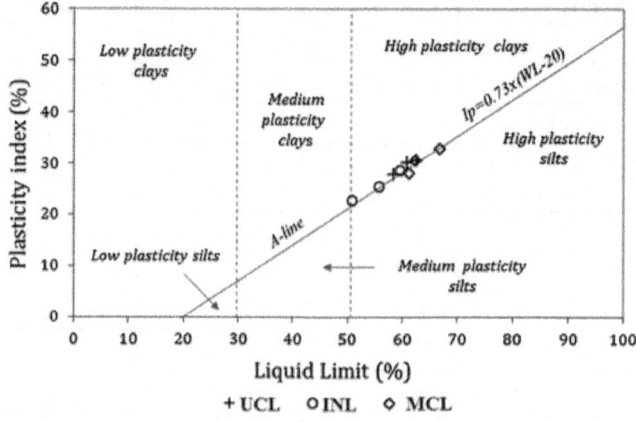

Figure 6: Casagrande chart classification of lateritic soils in the study area.

Satisfactory classification of expansive soils implies that the geotechnical parameters that characterize swelling are known (Millogo et al. 2008). The swelling rate (ε_s) which is defined as the percentage of swelling of a clay sample compacted at optimum Proctor and subjected to a load of 7 kPa was computed using the following equation (2) (Derriche and Kebaili 1998):

$$\varepsilon_s = 1.10^{-5} \times PI^{2.24}$$

(2)

where ε_s is the swelling rate and PI the plasticity index.

According to Djedid et al. (2001) and references therein, the lateritic soils that contain clay fraction between 8 and 65% are considered as low, moderate and high swelling soils if ε_s value range between 0–1.5, 1.5-5, and 5–25, respectively. The ε_s values presented in Table 6 shows that UCL and MCL soil samples (average of 2.09 and 2.12, respectively) fall within the high swelling soils range whereas the INL samples (average of 1.4) are of low swelling potential. Based on the classification proposed by Bekkouche et al. (2000) which defined the swelling potential by the relation between clay content and PI (Figure 7), the INL samples are plotted in the moderate swelling potential area while INL and MCL samples plot toward very high swelling potential area. This indicates that the UCL and MCL samples could be more susceptible to considerable change in volume in a changing climate.

Table 6: Estimated swelling potential of studied soils

Designation	Swelling potential (ε_s)			
	Nkil	Nkol	Mfi	Average
Upper clayed layer (UCL)	2.26	1.90	2.11	2.09
Intermediate nodular layer (INL)	1.80	1.40	1.09	1.43
Bottom clayed layer (MCL)	2.11	1.75	2.50	2.12

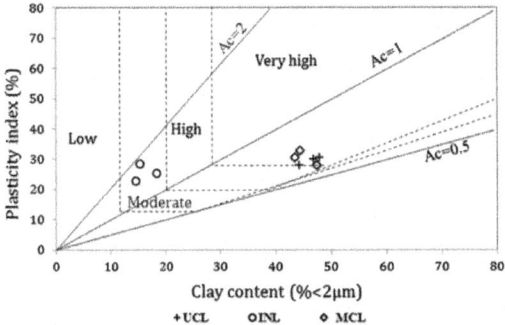

Figure 7: Domains of classification of soils according to their plasticity index (in Bekkouche et al., (2000)).Ac: activity of clays.

COMPACTION TEST

The modified Protor test variables and results of the statistical summary are presented in Table 3 and Table 4, respectively. The maximum dry density (MDD) values obtained for the investigated samples are in average 1.7, 2.1 and 1.8 g/cm³ for UCL, INL and MCL, respectively. While the optimum moisture content (OMC) values are relatively higher and almost similar for UCL and MCL soil samples (average of 18.5 and 17.9%), it is lower for INL soil samples (average of 12.2%). In the overall samples, it can be seen that MDD is higher when OMC is lower (Table 3). However, the discrepancy of these parameters in the samples may be due to the different degree of laterization in the weathering profiles.

According to O'Flaherty (1988) the range of values that may be anticipated when using the modified proctor test methods are: for clay, MDD may fall between 1.4 and 1.7 g/cm³ and optimum moisture content (OMC) may fall between 20-30%. For silty clay MDD is usually between 1.6 and 1.8 g/cm³ and OMC ranged between 15-25%. For sandy clay, MDD usually ranged between 1.8 and 2.2 g/cm³ and OMC between 8 and 15%. Thus, looking at the results in Table 3, it could be noticed that INL samples belong to sandy clay soils whereas UCL and MCL samples are rather silty clay soils. As per CEBTP specification (CEBTP, 1984), lateritic soils having MDD value of 2.0 and 1.8 g/cm³ are recommended to be used as raw materials in base and sub-base, respectively in road construction. Only INL samples are consistent with this specification and can be rated as excellent materials for road works.

Californian bearing ratio (CBR)

The overall soaked CBR results presented in Table 3 and 4 show both lateral and vertical variation with the values range between 15 and 54%. The soil samples collected at the site MFI shows CBR values higher than those collected at sites NKIL and NKOL (Table 3). Along the vertical transect, the CBR values are generally higher (with average of 39±19%) in the INL samples and lower (average of 19±3%) in MCL samples (Table 4). However, the study of the vertical profile of lateritic soils indicates that samples with lower proportions of fines (<80 μm) (INL) have higher values of MDD and CBR and vice versa. Thus, it can be deduced that the lower the fine components, PI and OMC, the higher the MDD and CBR values.

Among the different investigated layers, only INL soil samples (average of 39±19%) exceeded the CBR value of not less than 30% specified by the CEBTP for sub-base materials (Table 5). None of the tested sampled meet the CEBTP specification of not less than 80% for base materials. This suggest that

INL samples may be useful as sub-base materials whereas the engineering properties of UCL and MCL samples are of poor quality which do not allow their uses as row materials in road pavement. However, due to its relative high PI, LL and swelling potential (Figure 6 and Tables 5), these soils require mechanical, physical and/or chemical improvements with hydraulic binders such as quicklime, bitumen and/or cement prior to their uses as base materials for road works.

Direct shear strength

The shear strength is one of the most important engineering properties of a soil, because it is required whenever a structure is dependent on the soil's shearing resistance. For most of the geotechnical designs concerning foundations, earthworks and slope stability issues, the soils are required to withstand shearing stress along with compressive stress (Nakao and Fityus 2009). The test was conducted under saturated conditions. The shear rate was adopted to be 2 mm/min. The consolidation stress range of 100–300 kPa was considered. For a given material, shear vs. normal stresses at failure lay on a straight line in the τ- σ plane, named failure envelope. In soil mechanics, the failure envelope is traditionally known as Mohr-Coulomb failure criterion and expressed as (Eq. 3):

$$\tau = c' + \sigma \tan\psi$$

(3)

where τ and σ represent shear and normal stresses, respectively. c' is the cohesion of soil and ψ is the internal friction angle. This failure criterion is obtained from a series of direct shear tests where maximum or residual stress are plotted against normal stress and then a failure envelope is calculated. The results of direct shear test is presented in Table 7. Figure 8b presents a Mohr-Coulomb envelope obtained from direct shear test of soils in the study area. Each experimental point corresponds to the maximum stress that the soils can withstand without cracking or breaking. Slope and intersect of the linear envelope correspond to angle of internal friction (ψ) and cohesion (c'), respectively. Figure 8a indicate that the shear stress corresponding to a fixed normal stress increases initially until it reaches the peak strength, which then decreases gradually toward its residual strength. The observed behavior is ascribed to overconsolidated soils. The corresponding stress path responses illustrated in Figure 8b exhibit essentially linear failure envelopes with respect to the normal stress axis. A best-fit envelope drawn through the data points in Figure 8b results in shear strength parameters c' and ψ of 62 kPa and 27.2°, respectively, suggesting that any shear stress or slope above these values may cause soil failure.

Table 7: Summary of direct shear test performed on the sample from Mfou

Time(s)	Displace-ment (mm)	Stress: 1 bar		Stress: 2 bars		Stress: 3 bars	
		Load per unit area (kpa)	Shear stress (Kpa)	Load per unit area (Kpa)	Shear stress (Kpa)	Load per unit area (Kpa)	Shear stress (Kpa)
15"	0.25	2100	76	2900	106	3400	124
30"	0.50	2500	92	3400	125	4300	158
45"	0.75	2600	96	3700	137	4500	167
1'00"	1.00	2600	97	3900	146	4600	172
1'15"	1.25	2700	102	4000	151	4700	177
1'30"	1.50	2700	103	4100	156	4700	179
1'45"	1.75	2700	103	4200	161	4700	180
2'00"	2.00	2700	104	4200	162	4700	182
2'15"	2.25	2700	105	4200	164	5000	195
2'30"	2.50	2700	106	4200	165	5100	201
2'45"	2.75	2800	111	4200	167	5100	203
3'00"	3.00	2800	112	4200	168	5100	205
3'15"	3.25	2900	117	4300	174	**5100**	**206**
3'30"	3.50	2900	118	4500	184	5000	204
3'45"	3.75	2900	120	**4500**	**186**	5000	206
4'00"	4.00	2900	121	4400	183	5000	208
4'15"	4.25	**2900**	**122**	4400	185	4900	206
4'30"	4.50	2800	119	4400	187	4900	208
4'45"	4.75	2800	120	4400	189	4900	210
5'00"	5.00	2800	121	4400	191	4800	208
5'15"	5.25	2800	122	4400	192	4800	210
5'30"	5.50	2800	124	4400	194	4800	212
5'45"	5.75	2800	125	4400	196	4800	214
6'00"	6.00	2800	126	4400	199	4800	217
6'15"	6.25	2800	128	4400	201	4800	219
6'30"	6.50	2800	129	4300	198	4800	221

6'45"	6.75	2800	130	4300	200	4800	224
7'00"	7.00	2800	132	4300	202	4800	226
7'15"	7.25	2800	133	4300	205	4800	228
7'30"	7.50	2700	130	4300	207	4800	231
7'45"	7.75	2700	131	4300	209	4800	234

The bold values represent the maximum load of the testing soils and its corresponding shear stress.

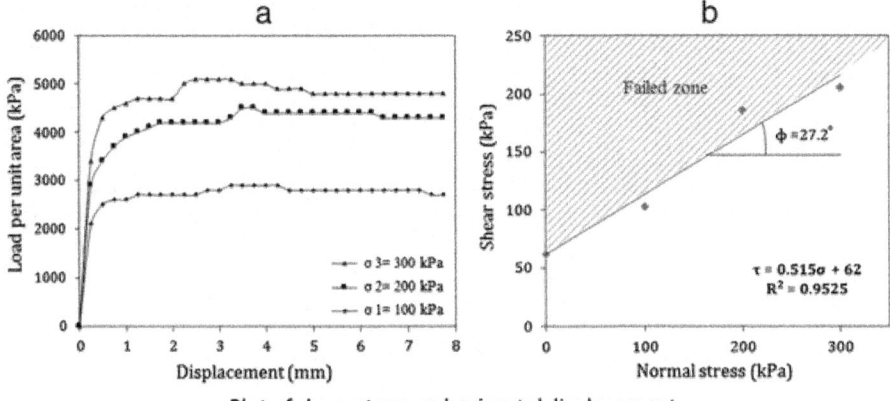

Plot of shear stress vs horizontal displacement

Figure 8: Stress path responses relative to direct shear strehgth for standard Protor campacted soils in the study area. (a) Plot of load per unit area vs horizontal displacement and (b) shear stress vs normal stress. The slope (62 Kpa) represents the cohesion and 27.2° represents the angle of the internal friction.

Consistent with the results reported in lateritic soils of Nigeria (cohesion range of 65–75 kPa and friction angle range between 26-31°) (Oladele et al. 2012; Aloa and Opaleye 2011), these values show that the soils in the study area have relative high bearing capacity as a result of their respective high cohesion and relatively medium angle of internal friction values. Furthermore, the plot of samples in the diagram of Dayre et al. (1978) modified by Ekodeck (1984) (Figure 9) shows that the samples have an elasto-plastic behavior. Hence, they can support slope stability and shallow foundation (Nakao and Fityus 2009).

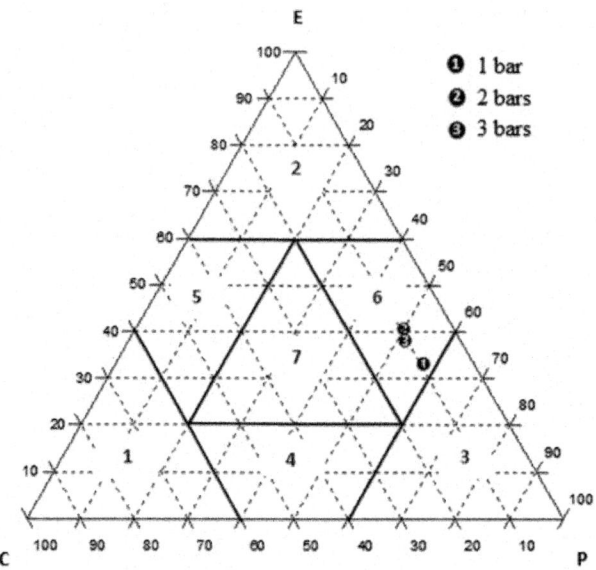

Figure 9: Characterization of soil samples based on the Dayre et al., (1978) diagram modified by Ekodeck, (1984). 1: compressibility, 2: elasticity, 3: plasticity, 4: plasto-compressibility, 5: elasto-compressibility, 6: elasto-plasticity, 7: elasto-plasto-compressibility.

Comparison of test results with some lateritic soils in Sub-Saharan Africa

To understand the regional variability of raw materials widely used in civil engineering works in Cameroon, the soils under consideration were compared with known lateritic soils in Sub-Saharan Africa. Globally, based on the CEBTP specification (CEBTP 1984) and as previously thought (Table 5), the engineering properties of these soils are all convenient to be used as sub-base materials and require pre-treatment before use as base materials (Table 8). However, significant differences are observed among their engineering parameters. The results show that there is an apparent similarity in parameters between soils formed under same climatic conditions. Despite their different parent rocks, lateritic soils under investigations have engineering properties relatively comparable with lateritic soils of Ghana, Ivory-Coast and Congo. Moreover, lateritic soils formed under semi-arid conditions have engineering properties higher than those formed under tropical and sub-tropical conditions. This is probably due to the intense weathering processes observed in tropical regions leading to removal of silica and accumulation of sesquioxides (IRRI1989). Several studies showed that silica has a positive correlation with specific gravity and CBR value. As an example, lateritic soils of Burkina Faso which are silica rich materials (SiO_2 of 56.19 wt.%) (Millogo et al.2008) have CBR value at 95% MDD (43%) higher than that in the study area (average CBR at 95% of 39) which are rather iron rich soils (Fe_2O_3 of 40.88 wt.%). It is worth mentioning that lateritic soils developed under sub-tropical conditions seemingly have the fairest engineering properties. Despite incomplete information regarding rock types, the results show that laterites developed on metamorphic rocks (schist, gneiss and quartzite) have engineering properties better than those developed on igneous rocks (granite).

Table 8: Comparison of studied lateritic soils with some lateritic soils in Sub-Saharan Africa

Country	Rock types	Climatic regime	Liquid limit (%)	Plastic limit (%)	Plasticity index (%)	Fines fraction (%)	OMC (%)	MDD	CBR (%)	AASHTO classification	References
Cameroon (INL)	Mi-cashist	Tropical	55	30	25	26	12	2.1	39	A-2-7	This study
Ghana	Granite	Tropical	-	-	-	30	12	2	-	-	Gidigasu, (1983)
Ivory Coast	Granite	Tropical	48	24	24	21	10	2.1	30	A-2-6	Bohi, (2008)
Congo	-	Tropical	80	-	40	35	19	1.7	30	A-2-7	Bohi, (2008)
Uganda	-	Sub-tropical	38	17	22	34	13	-	19	A-2-6	Zelalem, (2005)
Kenya	-	Sub-tropical	45	31	14	28	19	-	52	A-2-7	Zelalem, (2005)
Gambia	-	Sub-tropical	36	16	20	22			-	A-2-6	Zelalem, (2005)
Tanzania	-	Sub-tropical	34	19	10	25	9	2.1	22	A-2-4	Zelalem, (2005)
Ethiopia	Granite, Schist	Sub-tropical	68	33	32	58	28	1.7	15	A-7-5	Zelalem, (2005)
Senegal	Schist-	Semi-arid	25	16	16	20	9	1.9	86	A-2-7	Bohi, (2008)
Niger	-	Semi-arid	21	11	10	25	9	2.1	22	A-2-4	Zelalem, (2005)
Nigeria	Gneiss, Quartzite	Semi-arid	30	16	14	52	18	2.5	60	A-2-6	Bello et Adegoke, (2010)
Burkina-Faso	Granite	Semi-arid	22	12	10	11	7	2.2	43	A-2-7	Millogo et al., (2008)
Sudan	-	Arid	21	12	9	27	-	-	-	A-2-4	Zelalem, (2005)

CONCLUSION

This research collates the baseline geotechnical information on tropical soils developed on micashist of Central-Cameroon and compares their geotechnical properties with some lateritic soils in Sub-Saharan Africa. Investigations on the lateral and vertical variability of the engineering properties of lateritic soils have helped in deducing the following conclusions:

The texture of the studied micaschist is lepidoblastic, granoblastic and porphyroblastic with variable size of minerals. The mineral assemblages and their modal composition are of quartz (35%), muscovite (25%), biotite (15%), garnet (10%), chlorite (6%), sillimanite (3%), plagioclase (3%) and K-feldspar.

The micashist derived lateritic soils from Mfou have a pigmentation varying from yellowish to reddish and were formed in an oxidizing environment having good drainage.

The mineralogical investigations indicate the absence of high swelling clays such as smectite and montmorillonite. Chemical test shows that these soils are "true laterites" having high concentration of sesqueoxides (iron and aluminum oxides) which may act as cementing agents contributing to tensile cracking by making the compacted structure relatively brittle.

The results of geotechnical tests reveal that samples from upper clay layer and bottom mottled clay layer are poorly graded soils with engineering properties (fines of 61 and 63%, plasticity index of 30 and 31%, CBR at 95% of 21 and 19%) which do not allow their use as raw materials in road construction. INL samples are gravel and sand soils well graded having engineering properties (fines of 26%, plasticity index of 26% and CBR of 39%) that meet specifications require for sub-base materials for light traffic roads. However, Atterberg's consistency limit tests indicate that the samples are moderate to high plastic soils with moderate swelling potential. Hence, their use as base and sub-base construction materials for heavy traffic roads require pre-treatment with non-plastic soils, bitumen, cement and/or lime. The direct shear test results show that the soils have high bearing capacity (cohesion of 62 kPa and 27.2° of angle of internal friction) making them to be useful in slope stability and in shallow foundation design.

The comparison of the studied soils with some lateritic soils in Sub-Saharan Africa indicates that genesis and climatic conditions are potentials factors that influence the engineering properties of lateritic soils. Further, the comparison also indicates that soils developed in semi-arid conditions exhibits the best engineering characteristics.

ACKNOWLEDGEMENTS

The work reported in this paper is part of on-going Ph.D research of the main author at the Graduate School of Engineering for Education, University of Toyama, Toyama City, Japan. Professor Georges Emmanuel Ekodeck and the personal of the National Laboratory of Civil Engineering (LABOGENIE) of Cameroon are gratefully acknowledged for Laboratory facilities. The authors wish to thank Professor Katsumi Marumo of the Departement of Environmental Biology and Chemistry, Faculty of Science, University of Toyama for his support to performing chemical analyses (X-ray fluorescence). An additional support was provided by the SATREPS-Nymo project. We also wish to express our gratitude to Mme. Emilia Bi Fantong, Mlle Edwige R. Tiodjio and for Dr. Jude M. Wirmvem for their vision, support and encouragement throughout the work. This paper has benefited from valuable contributions from two anonymous reviewers and we appreciate their efforts.

COMPETING INTERESTS

The authors declare that they have no competing interests.

REFERENCES

1. AFNOR (Association française de normalisation). (1978). NF P18-560, Analyse granulométrique par tamisage(p. 5).

2. AFNOR (Association française de normalisation). (1992). NF P94-057, Analyse granulométrique des sols, Méthode par sédimentation (p. 17).

3. AFNOR (Association française de normalisation). (1993). NF P94-051, Détermination des limites d'Atterberg (p. 15).

4. AFNOR (Association française de normalisation). (1997). NF P94-078, Indice CBR après immersion - Indice CBR immédiat - Indice Portant Immédiat (p. 14).

5. AFNOR (Association française de normalisation). (1999). NF P94-093, Détermination des références de compactage d'un matériau, NF P94-071-2, Essai de cisaillement direct-principe et méthode (p. 17).

6. Alao, D. A., & Opaleye, S. T. (2011). Geotechnical analysis of slope failure of a kaolin Quarry at Kura, Jos North-Central. Int J Sci Res, 1(1), 87–102.

7. Autret, P. (1983). Latérites et graveleux latéritiques. Laboratoire Central des Ponts et Chaussées (p. 38).

8. Bagarre, E. (1990). Utilisation des graveleux latéritiques en technique routière (p. 143). Paris: Institut des Sciences et des Techniques de

l'Equipement et de l'Environnement.

9.   Bayewu, O. O., Olountola, M. O., Mosuro, G. O., & Adeniyi, S. A. (2012). Petrographic and geotechnical properties of Lateritic Soils developed over different parent rocks in Ago-Iwoye area, Southwestern Nigeria. Int J of Appl Sci and Eng Res, 4, 584–594.

10.   Bekkouche, A., Djedid, A., & Aissa Mamoune, S. M. (2000). An experimental investigation on the assessment of the swelling parameters, IV. Int Congr East Mediterr Univ Gazimagusa North Cyprus, 2, 627–633.

11.   Bello, A. A., & Adegoke, C. W. (2010). Evaluation of geotechnical properties of Ilesha East Southwestern Nigeria's Lateritic Soil. Pac J Sci Technol, 11, 617–624.

12.   Bohi, Z. H. B. (2008). Caractérisation des sols latéritiques utilisés en construction routière. Cas de la région de l'Agnéby (Côte d'Ivoire). Thèse de Doctorat (p. 143). Paris: École Nationale des Ponts et Chaussées.

13.   Bouassida, R., & Boussetta, R. (2007). Manuel de travaux pratiques de mécanique des sols (p. 116). Tunis, Tunisie: Centre de publication universitaire.

14.   CEBTP (Centre Expérimental de Recherches et d'Etudes du Bâtiment et des travaux Publics). (1984). Guide pratique du dimensionnement des routes pour les pays tropicaux (p. 157).

15.   Day, R. W. (2000). Geotechnical Engineer's Portable Handbook. New York: McGraw-Hill.

16.   Dayre, J., Fabre, D., Letourneur, J., Antoine, P., & Orengo, Y. (1978). Eléments pour une classification géotechnique des terrains (2nd ed., pp. 131–139). Madrid II: C.R. 3e Cong. AIGI.

17.   De Graft-Johnson, J. W. S. (1972). Lateritic gravel evaluation of road construction. J soil Mech Div Amst Soc Civil Eng, 98, 1245–1265.

18.   Deckers, J., Nachtergaele, F., & Spaargaren, O. (2001). Tropical Soils in the Classification Systems of USDA, FAO and WRB, KULeuven, Institute for Land and Water Management, Belgium. The Netherlands: Land and Water Development Division, International Soil Reference and Information Centre (ISRIC).

19.   Derriche, Z., & Kebaili, M. (1998). Prévision du gonflement des argiles d'In-Aménas. Bulletin des Laboratoires des Ponts et Chaussées, 218, 15–23.

20.   Djedid, A., Bekkouche, A., & Aissa Mamoune, M. (2001). Identification and prediction of the swelling behavior of some soils from the Tlemcen region of Algeria. Bulletin des Laboratoires des Ponts et Chaussées, 233,

69–77.

21. Ekodeck, G. E. (1984). L'altération des roches métamorphiques du Sud Cameroun et ses aspects géotechniques. Thèse de Doctorat D'Etat des Sciences Naturelles, IRGM (p. 368). France: Université Scientifique et Médicinal de Grenoble I.

22. Ekodeck, G. E. (1985). Implications géotechniques de l'altération des roches de la région de Yaoundé. Rev Sci et Tech série Science de la Terre, 1, 55–82.

23. Ferry, J. M., & Spear, F. S. (1978). Experimental calibration of the partitioning of Fe and Mg between biotite and garnet. Contrib Mineral Petrol, 66, 13–117.

24. Garg, S. K. (2009). Soil Mechanics and Foundation. Engineering (7th ed., pp. 673–683). New Delhi: Khana publishers.

25. Gazel, J., & Guiraudie. (1967). Carte géologique de reconnaissance du Cameroun au 1/500 000, feuille Yaoundé Est, N° NA 32 NE-E22 (5th ed., p. 29). Du Cameroon: Bul. Dir. Mines & Géol.

26. Gidigasu, M. D. (1983). Development of acceptance specifications for tropical gravel paving materials. Eng Geo, 19, 213–240.

27. Indraratna, B., & Nutalaya, P. (1991). Engineering characteristics of a compacted, lateritic residual soil.Geotech Geol Eng, 9, 125–137.

28. IRRI (International Rice Research Institute). (1989). Climate and Food Security (p. 602). Manila, Philippines: □.

29. Kabir, M. H., & Taha, M. R. (2004). Assessment of physical properties of a granite residual soil as an isolation barrier. Electron J Geotech Eng, 92, 13.

30. Kamgang Beyala, V. (1987). L'altération supergène des roches grenatifères de la région de Yaoundé (Cameroun): Pétrologie-Minéralogie. Thèse de Doctorat 3e Cycle (p. 170). France: Université de Poitiers.

31. Leake, B. E., Hendry, G. L., Kemp, A., Plant, A. G., Harvey, P. K., Wilson, J. R., Coats, J. S., Aucott, J. W., Lunel, T., & Howarth, RJ. (1970). The chemical analysis of rock powders by automatic X-ray fluorescence. Chem Geol, 5, 7–86.

32. Lemougna, P. N., Chinge Melo, U. F., Kamseu, E., & Tchamba, A. B. (2011). Laterite based stabilized products for sustainable building applications in tropical countries: review and prospects for the case of Cameroon. Sustainability, 3, 293–305. doi: 10.3390/su3010293.

33. Logmo, E. O., Ngon Ngon, G. F., Samba, W., Mbog, M. B., & Etame, J. (2013). Geotechnical, mineralogical and chemical characterization

of the missole II clayey materials of Douala Sub-Basin (Cameroon) for construction materials. J Civ Eng. , ▨. doi: 10.4236/ojce.2013.32A006.

34. Mahalinger-Iyer, U., & Williams, D. J. (1997). Properties and performance of lateritic soil in road pavements.Eng Geol, 46, 71–80.

35. Mbumbia, L., De Wilmarsa, A. M., & Tirlocp, J. (2000). Performance of lateritic soil bricks fired at low temperatures: a case study of Cameroon. Construct Build Mater, 14, 121–131.

36. Millard, R. S. (1993). Road Building in the Tropics (9th ed., p. 312). London, UK: Transport Research Laboratory, State of the Art Review, Hazardous Materials Storage Ordinance Publications.

37. Millogo, Y., Traore, K., Ouedraogo, R., Kabore, K., Blanchart, P., & Thomassin, J. H. (2008). Geotechnical, mechanical, chemical and mineralogical characterization of a lateritic gravels of Sapouy (Burkina Faso) used in road construction. Const Build Mat, 22, 70–76.

38. Molua, EL & Lambi, CM (2006). Climate hydrology and water resources in Cameroon CEEPA, Pretoria, p 37 ☐ .

39. Munsell color charts. (1975). Color Guides (p. 34). Baltimore Md: Handbook manual.

40. Nakao, T., & Fityus, S. (2009). Direct shear testing of a marginal material using a large shear box. Geotech Test J, 31, 5. ID GTJ101237.

41. Naresh, C. S., & Nowatzki, P. E. (2006). Soils and foundations. Ref Man Fed Highway Adm, 2, 8–75.

42. Ndam, N. J. R., Braun, J. J., Meybeck, M., & Bedimo Bedimo, J. P. (1998). Réactualisation des données hydroclimatiques des bassins fluviaux de la Sanaga et du Nyong (Sud Cameroun). Presse Universitaire de Yaoundé, 2, 51–64.

43. Northmore, K. J., Culshaw, M. G., Hobbs, P. R., Hallam, J. R., & Entwisle, D. C. (1992). Engineering Geology of Tropical Soils – Summary Finding and their Application for Engineering Purposes. Nottingham: Technical report WN/93/15, Overseas Development Administration (ODA) and British Geological Survey (BGS.

44. Nzenti, J. P., Barbey, P., Macaudière, J., & Soba, D. (1988). Origin and evolution of the late Precambrian high-grade Yaoundé gneisses (Cameroon). Precam Res, 38, 91–109.

45. O'Flaherty, C. A. (1988). Highway Engineering (Vol. 2). London, UK: Edward Arnold Publishers.

46. Ojo, J. S., & Adeyemi, G. O. (2003). Opportunities for ventures in construction materials. In A. A. Elueze (Ed.),Prospects for Investment in

Mineral Resources of Southwestern Nigeria (pp. 47–54).

47. Oladele, A. O., Olusolo, J. O., & Emmanuel, T. A. (2012). Engineering properties of lateritic soils around Dall Quarry in Sango area, Ilorin, Nigeria. Earth Sci Res, 1, 2. doi: 105539/esr.v1n2p71.

48. Onana, V. L. (2010). Les schistes de la série de Mbalmayo-Bengbis et leurs produits dérivés: Caractérisations altérologiques et géotechniques. Thèse de Doctorat Ph.D (p. 236). Yaoundé: Université de Yaoundé I.

49. Que, J., Wang, Q., Chen, J., Shi, B., & Meng, Q. (2008). Geotechnical properties of the soft soil in Guangzhou College City. Bull Eng Geol Environ, 67, 479–483.

50. Ramamurthy, T. N., & Sitharam, T. G. (2005). Geotechnical Engineering (p. 289). New Delhi: S. Chand.

51. Robitaille, V., & Tremblay, D. (1997). Mécanique des sols (théorie et pratique) (p. 652). Quebec-Canada: Modulo.

52. Rossiter, D.G., 2004. Digital soil resource inventories: status and prospects. Soil Use & Management 20(3)296–301

53. Segalen, P. (1967). Les sols et la géomorphologie du Cameroun. ORSTOM, 2, 137–187.

54. Sikali, F., & Mir-Emarati, D. (1986). Utilisation des latérites en technique routière au Cameroun. In Séminaire Régional sur les Latérites: Sols, Matériaux, Minerais, Douala (CMR) (pp. 277–288). ISBN 2-7099-0867-0.

55. Tardy, Y. (1992). Pétrologie des latérites et sols tropicaux. Masson (ed) (p. 459).

56. Tockol, I., Massiéra, M., Chiasson, P. A., & Maiga, M. S. (1994). Les graveleux latéritiques dans les pays du sahel: Cas des routes non revêtues (pp. 3423–3431). Rotterdam, Holland: 7ème Congrès International de l'AIGI.

57. TPSSA (Transport Policies in Subsaharienne Africa). (2004). Afrique-Cameroun-Road, Report (p. 54).

58. Tsozué, D., Bitom, D., & Yongue-Fouateu, R. (2012). Morphology, mineralogy and geochemistry of a lateritic sequence developed on micaschist in the Abong-Bang region, Southeast Cameroon. Geological Soc S Afr, 115, 103–116.

59. Vallerie, M. (1995). ATLAS Régional du Sud-Cameroun (p. 70). Paris: ORSTOM.

60. Yongue, F. R. (1986). Contribution à l'étude pétrologique de l'altération et des faciès de cuirassement ferrugineux des gneiss migmatitiques de

la région de Yaoundé (p. p 214). Yaoundé: Thèse, Doctorat 3e cycle Univisité de Yaoundé.

61. Zelalem, A. (2005). Basic Engineering Properties of Lateritic Soils Found in Nejo-Mendi Road Construction Area, Welega, M. Sc. thesis (p. 97). Ethiopia: Addis Ababa University.

CITATION

CHAPTER 1

Santosh Kumar Sarkar, Paulo J.C. Favas, Dibyendu Rakshit and K.K.Satpathy (2014). Geochemical Speciation and Risk Assessment of Heavy Metals in Soils and Sediments, Environmental Risk Assessment of Soil Contamination, Dr. Maria C. Hernandez Soriano (Ed.), ISBN: 978-953-51-1235-8, InTech, DOI: 10.5772/57295.

CHAPTER 2

Wisley Moreira Farias, Geraldo Resende Boaventura, Éder de Souza Martins, Fabrício Bueno da Fonseca Cardoso, José Camapum de Carvalho and Edi Mendes Guimarães (2014). Chemical and Hydraulic Behavior of a Tropical Soil Compacted Submitted to the Flow of Gasoline Hydrocarbons, Environmental Risk Assessment of Soil Contamination, Dr. Maria C. Hernandez Soriano (Ed.), ISBN: 978-953-51-1235-8, InTech, DOI: 10.5772/57234.

CHAPTER 3

A.K.M. Azad Hossain and Greg Easson, Soil Moisture Estimation in South-Eastern New Mexico Using High Resolution Synthetic Aperture Radar (SAR) Data, doi:10.3390/geosciences6010001.

CHAPTER 4

Dale W. Griffin, Erin E. Silvestri, Charlena Y. Bowling, Timothy Boe, David B. Smith and Tonya L. Nichols, Anthrax and the Geochemistry of Soils in the Contiguous United States, doi:10.3390/geosciences4030114.

CHAPTER 5

Mohamed A. Shahin, Mark B. Jaksa, and Holger R. Maier, "Recent Advances and Future Challenges for Artificial Neural Systems in Geotechnical Engineering Applications," Advances in Artificial Neural Systems, vol. 2009, Article ID 308239, 9 pages, 2009. doi:10.1155/2009/308239.

CHAPTER 6

Isaac A. Jeldes, Eric C. Drumm , John S. Schwartz; The Low Compaction Grading Technique on Steep Reclaimed Slopes: Soil Characterization and Static Slope Stability; DOI 10.1007/s10706-013-9648-0

CHAPTER 7

Luı́s Pinheiro Branco,Antoʹnio Topa Gomes,Antoʹnio Silva Cardoso,Carla Santos Pereira; Natural Variability of Shear Strength in a Granite Residual Soil from Porto; DOI 10.1007/s10706-014-9768-1

CHAPTER 8

Hamed A. Keykha,Bujang B. K. Huat,Afshin Asad; Electrokinetic Stabilization of Soft Soil Using CarbonateProducing Bacteria; DOI 10.1007/s10706-014-9753-8

CHAPTER 9

Mohammed Y. Fattah; Hawraa H. M. al-Musawi, Firas A. Salman; Treatment of Collapsibility of Gypseous Soils by Dynamic Compaction; DOI 10.1007/s10706-012-9552-z

CHAPTER 10

Brendan C. O'Kelly; Compression and consolidation anisotropy of some soft soils; DOI 10.1007/s10706-005-5760-0

CHAPTER 11

Brice T Kamtchueng, Vincent L Onana , Wilson Y Fantong , Akira Ueda1 , Roger FD Ntouala , Michel HD Wongolo , Ghislain B Ndongo , Arnaud Ngo'o Ze , Véronique KB Kamgang and Joseph M Ondoa; Geotechnical, chemical and mineralogical evaluation of lateritic soils in humid tropical area (Mfou, Central-Cameroon): Implications for road construction; DOI 10.1186/s40703-014-0001-0

INDEX